Myroslav Kryven

Optimizing Crossings
in Circular-Arc Drawings and Circular Layouts

Myroslav Kryven

Optimizing Crossings in Circular-Arc Drawings and Circular Layouts

Würzburg
University Press

Dissertation, Julius-Maximilians-Universität Würzburg
Fakultät für Mathematik und Informatik, 2020
Gutachter: Prof. Dr. Alexander Wolff, Prof. Dr. André Schulz

Impressum

Julius-Maximilians-Universität Würzburg
Würzburg University Press
Universitätsbibliothek Würzburg
Am Hubland
D-97074 Würzburg
www.wup.uni-wuerzburg.de

© 2022 Würzburg University Press
Print on Demand

Coverabbildung: Myroslav Kryven
Coverdesign: Jule Petzold

ISBN 978-3-95826-174-7 (print)
ISBN 978-3-95826-175-4 (online)
DOI 10.25972/WUP-978-3-95826-175-4
URN urn:nbn:de:bvb:20-opus-245960

Zusammenfassung

Ein Graph ist eine Datenstruktur bestehend aus einer Menge von Objekten (die Knoten genannt werden) und einer Menge von Beziehungen (die Kanten genannt werden) zwischen Paaren von Objekten. Graphen modellieren verschiedene Arten von Netzwerken. Zum Beispiel entsprechen in sozialen Netzwerken die Knoten Personen und die Kanten stellen ihre Freundschaftsbeziehungen dar. Weitere Beispiele sind neurologische Netzwerke, Straßennetze, UML-Diagramme, die alle durch Graphen dargestellt werden können. Um verschiedene Probleme auf solchen Netzwerken zu lösen, können häufig Standard-Graphalgorithmen verwendet werden. Ein Navigationssystem kann zum Beispiel eine Route zwischen zwei Orten auf einer Karte finden, indem es einen kürzesten Weg zwischen zwei Knoten im Graphen berechnet.

Um die Struktur eines Graphen zu verdeutlichen, ist es hilfreich den Graphen zu visualisieren. Das Forschungsgebiet der Visualisierung von Graphen heißt *Graphenzeichnen*. Es gibt viele Möglichkeiten, wie ein Graph visualisiert werden kann. Eine klassische Visualisierungsmethode für Graphen sind sogenannte *Node-Link-Diagramme*. Bei dieser Darstellung werden die Knoten als Punkte gezeichnet und für jedes Paar von Knoten, die im Graph benachbart sind, werden die entsprechenden Punkte durch eine Kurve verbunden. Die Kanten können durch Strecken, Polygonzüge, Kreisbögen oder allgemeine Jordankurven repräsentiert werden.

Bei solchen Darstellungen möchte man Kreuzungen zwischen Kanten vermeiden, weil Kreuzungen die Lesbarkeit einer Zeichnung verringern. Graphen, die ohne Kreuzungen gezeichnet werden können, heißen *planare* Graphen. Planare Graphen sind intensiv untersucht worden. Zum Beispiel können planare Graphen effizient erkannt und auf einem kleinen Gitter gezeichnet werden. Wenn hingegen viele Kreuzungen nötig sind, um einen Graphen zu zeichnen, gibt es wenig Hoffnung auf eine lesbare Zeichnung. Darüber hinaus gibt es, in vielen Fällen, für solche Graphen keine effizienten Algorithmen, um sie zu erkennen oder zu zeichnen. Deswegen ist *Kreuzungsminimierung* ein fundamentales Thema im Graphenzeichnen. Graphen, die mit wenig Kreuzungen gezeichnet werden können, heißen *beyond-planar*. Das Thema, das sich mit Definition und Analyse von beyond-planaren Graphen beschäftigt, heißt *Beyond Planarity* und ist ein wichtiges, noch recht junges Forschungsgebiet im Graphenzeichnen.

Generell gilt für beyond-planare Graphen, dass sie eine Zeichnung besitzen, bei der die Art der Kreuzungen irgendwie eingeschränkt ist; zum Beispiel, wenn die Anzahl der Kreuzungen durch eine Konstante beschränkt ist (unabhängig von der Größe des Graphen). Kreuzungen können auch lokal beschränkt werden, indem wir zum Beispiel höchstens eine konstante Anzahl von Kreuzungen pro Kante erlauben oder höchstens eine konstante Anzahl von sich paarweise kreuzenden Kanten erlauben. Kreuzungen können auch dadurch beschränkt werden, dass wir den Winkel, unter dem sich kreuzende Kanten schneiden, nach unten beschränken. Diese Dissertation beschäftigt sich mit Klassen von beyond-planaren Graphen, die durch solche lokalen Einschränkungen von Kreuzungen definiert sind.

Contents

Contents

Introduction

A *graph* is an abstract network that represents a set of objects, called *vertices*, and relations between these objects, called *edges*. Graphs can model various networks. For example, a social network where the vertices correspond to users of the network and the edges represent relations between the users. Neural networks, road networks, integrated circuit networks, UML diagrams can all be modeled as graphs. Graph algorithms can be used then to efficiently solve a wide range of problems on these networks. For example, a navigation system can compute a route between two locations by finding a shortest path in the corresponding graph that models the road network.

Graph drawing.   Essentially a graph is just a data structure consisting of vertices and edges between these vertices. For a human to get more insight into the structure of the graph it is sometimes helpful to represent it visually. The field of visualizing graphs is called *Graph Drawing*. There are many ways to visualize a graph. A standard visualization is a *node-link diagram* in the Euclidean plane. In such a representation the vertices are drawn as points, or possibly other geometric objects like disks or polygons, in the plane and edges are drawn as Jordan curves between every two vertices connected by an edge. The Jordan curves used for edges can be straight-line segments, polygonal lines, circular-arcs, or piece-wise smooth curves. Another classical visualization is a *contact representation* where vertices are represented as geometric objects in the Euclidean plane (disks, polygons, as well as 1-dimensional objects like Jordan curves) such that for any two objects representing two vertices connected by an edge in the graph these objects must have a non-empty intersection in the representation. The various choices that influence the appearance of the drawing, vertices, or edges are usually referred to as the *drawing style*.

A *readable* drawing is a drawing that can be easily visually perceived by a human eye. In a readable drawing, simple tasks can be answered quickly by a human, for example, tracing an edge or a path between two vertices, finding a node with the largest degree etc.

Various measures for readability of a drawing have been studied such as the *angular resolution* (that is, the minimum angle between any two adjacent edges), the *number of crossings* in a drawing, *crossing angle resolution* (that is, the minimum crossing angle in the drawing), the *area* of the drawing etc. Such measures are called *quality measures* of a drawing. It is one of the main goals of Graph Drawing to define and study quality measures to produce readable drawings of graphs.

Beyond planarity.   A crossing between two edges can make it difficult for a user to trace an edge, as the user might accidentally jump to another edge while tracing one. Thus, crossings might adversely affect the readability of a drawing. This hypothesis was confirmed by a series of cognitive experimental studies [Pur00, PCA02, WPCM02]. Thus, ideally we would like to draw graphs without crossings. Graphs that can be drawn without crossings are known in literature as *planar* graphs. However, many graphs cannot be drawn without crossings, for example, those that have more than $3n - 6$ edges, where n is the number of vertices. Therefore, it is a natural problem to find drawings of graphs with as few crossings as possible or drawings where crossings affect the readability as little as possible. This problem is called *crossing optimization*.

Some graphs require a lot of crossings to be drawn [ACNS82] which makes it difficult to find readable drawings for these graphs. Moreover, a lot of graph problems that can be solved efficiently on planar graphs become intractable on general graphs. Thus, it is natural to consider graphs that are "close" to planar graphs, that is, graphs that admit drawings with restrictions on crossings, these graphs are known as *beyond-planar graphs*. One type of such restrictions can be on a crossing pattern, for example, graphs that admit a drawing where there are no edge intersected more than k times, k pairwise crossing edges, two adjacent edges that cross a third one etc. Another known type of restrictions is on the crossing angle resolution of the drawing of a graph, for example, by requiring it to be 90°. Studying graphs that admit drawings with such restrictions on crossings is the aim and scope of the area of Graph Drawing *Beyond Planarity*. For a more comprehensive introduction to the topic see the survey by Didimo et al. [DLM18].

1.1 Outline of the Book

This book consists of two parts. In part Part I we study Crossing Optimization in circular-arc drawings. First we consider crossing-free circular-arc drawings of graphs in Chapter 3. Then in Chapter 4 we introduce right-angle crossings between circular arcs by studying properties of circles that intersect at right angles only. Finally, in Chapter 5, we turn to graphs that can be drawn with circular arcs and right-angle crossings only. Part I consists of three peer reviewed papers.

In Part II we study Crossing Optimization in circular layouts. In Chapter 6 we consider edge crossing minimization and then in Chapter 7 we turn our attention to bundled crossing minimization in circular layouts. Part II consists of two peer reviewed papers.

1.1.1 Optimizing Crossings in Circular-Arc Drawings

The idea of drawing graphs with circular arcs dates back to at least the work of the artist Mark Lombardi who drew social networks, featuring players from the political and financial sector [LH03]. Indeed, user studies [PHNK13, XRPH12] state that users prefer edges drawn with curves of small curvature; not necessarily for performance but for aesthetics. Moreover, drawing graphs with circular arcs can help to improve certain quality measures of a drawing. Cheng et al. [CDGK01] showed, in particular, that using C^1-continuous curves consisting of at most three circular arcs, a graph can be drawn in small area and with optimal angular resolution. Aichholzer [AAA$^+$12] also studied angular resolution optimization in circular-arc drawings of triangulations. Recently, a new type of quality measure was introduced: the number of geometric objects that are needed to draw a graph given a certain style; note that two edges can belong to the same geometric object. Schulz [Sch15] coined this measure the

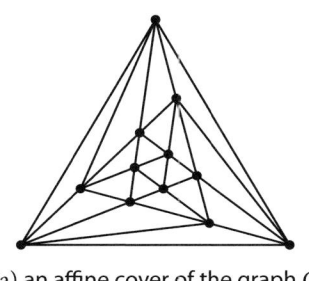

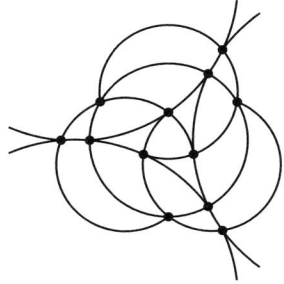

(a) an affine cover of the graph G (b) a spherical cover of the graph G

Figure 1.1: Affine and spherical covers of the same graph I (the icosahedron). Note, that the drawing (a) is segment optimal and the drawing (b) is circular-arc optimal; see Chapter 3.

visual complexity of a drawing and provided algorithms yielding circular-arc drawings with better visual complexity than those known for straight-line drawings.

Covering Graphs with Few Circles and Few Spheres

When optimizing crossings in Graph Drawing, the first question that we might ask ourselves is "Can the given graph be drawn without crossings?". So if the graph is planar, we should also aim for a planar drawing. If the graph is not planar it still can be drawn without crossings in 3D but this might come at some other expense. Chaplick et al. [CFL+16] introduced a quality measure for the visual complexity, the *affine cover number*, which is the minimum number of lines (or planes) that together cover a crossing-free straight-line drawing of a graph G in 2D (3D). Note that this quality measure considers crossing-free drawings of graphs, in particular, in 2D it is defined only for planar graphs, whereas in 3D it is also defined for non-planar graphs. For non-planar graphs, however, it is natural that the affine cover number increases with the density of a graph as well as with the number of crossings that a graph needs to be drawn in a single plane. Therefore, drawing graphs in 3D may reduce the number of crossings but increase the visual complexity.

In Chapter 3, we introduce the *spherical cover number*, which is the minimum number of circles (or spheres) that together cover a crossing-free circular-arc drawing in 2D (or 3D). (See for an example of an affine cover and a spherical cover of the same graph Figures 1.1a and 1.1b, respectively.) It turns out that spherical covers are sometimes significantly smaller than affine covers. For graphs from certain graph classes, we analyze their spherical cover numbers and compare them to their affine cover numbers. We also link the spherical cover number to other graph parameters (which we define in Section 2.1) such as *treewidth* and *linear arboricity*.

This chapter is based on joint work with Alexander Ravsky and Alexander Wolff [KRW19].

On Arrangements of Orthogonal Circles

In Chapter 4, we analyze properties of *orthogonal* circles, that is, circles that intersect at right angles only as well as properties of arrangements of such circles, that is, arrangements of circles

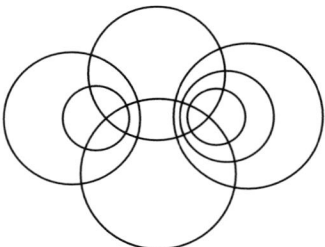

Figure 1.2: Apollonian circles.

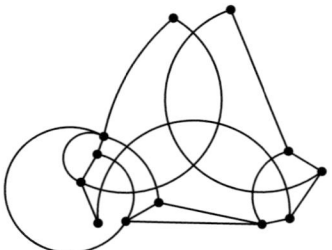

Figure 1.3: An arc-RAC drawing.

where every pair of circles must either be disjoint or orthogonal. A classical arrangement of orthogonal circles based on Apollonian circles (see Chapter 4 for definition) is illustrated in Figure 1.2. Using geometric arguments, we show that arrangements of orthogonal circles have only a linear number of intersections. This implies that *orthogonal circle intersection graphs* (that is, the graphs that have a vertex for each circle and an edge for each pair of orthogonal circles) have only a linear number of edges. When we restrict ourselves to orthogonal *unit* circles, the resulting class of intersection graphs is a subclass of *penny graphs* (that is, contact graphs of unit circles). We show that, similarly to penny graphs, recognizing orthogonal unit circle intersection graphs is as hard as many other problems for which no efficient algorithm is known. Such problems are called *NP-hard*. We define the complexity class of such problems more formally in Section 2.3.

This chapter is based on joint work with Steven Chaplick, Henry Förster, and Alexander Wolff [CFKW19].

Drawing Graphs with Circular Arcs and Right-Angle Crossings

In contrast to Chapter 3, in Chapter 5 we turn to drawing graphs with circular arcs allowing crossings, but only with *optimal* crossing angle resolution, that is, we insist that any two crossing edges must cross at a right angle.

In a *RAC drawing* [DEL11] of a graph, vertices are represented by points in the plane, adjacent vertices are connected by line segments, and crossings must form right angles. Graphs that admit such drawings are called *RAC graphs*. RAC graphs are beyond-planar graphs and have been studied extensively. In particular, it is known that a RAC graph with n vertices has at most $4n - 10$ edges. We introduce a superclass of RAC graphs, which we call *arc-RAC graphs*. In an arc-RAC drawing, edges are drawn as circular arcs (we consider a straight-line segment to be a circular-arc of infinite radius) and crossings must still form right angles; see Figure 1.3. We prove that an arc-RAC graph with n vertices has at most $14n - 12$ edges and that there are n-vertex arc-RAC graphs with $4.5n - O(\sqrt{n})$ edges.

This chapter is based on joint work with Steven Chaplick, Henry Förster, and Alexander Wolff [CFKW20].

1.1.2   Optimizing Crossings in Circular Layouts

Crossing minimization is a fundamental problem in graph drawing. Like many problems in Graph Drawing, crossing minimization in general graphs is NP-hard [GJ83]. It also remains NP-hard under some restrictions [Hli06]. But plenty of variants of the problem are known. The minimum number of crossings that a graph can be drawn with is known as the *crossing number* of the graph. In his seminal survey [Sch17] Schaefer recorded at least 89 different notions of crossing numbers. In Part II we deal with some of them. We drop the restriction that edges are drawn as circular arcs but insist instead on *circular layouts* of graphs, that is, that the vertices lie in convex position, for example on a circle, and the edges are drawn inside the disk of this circle. Such drawings are also known as *convex drawings*.

Edge Crossing Minimization in Circular Layouts

In Chapter 6 we consider edge crossing minimization in circular layouts, in particular, we study the following two beyond-planar graph classes:

- *outer k-planar graphs*, that is, graphs that admit a circular layout where each edge is crossed by at most k other edges; and

- *outer k-quasi-planar graphs*, that is, graphs that admit a circular layout where no k edges cross pairwise;

see for example in Figure 1.4a a drawing which is outer 3-quasi-planar but not outer 12-planar.

We show that outer k-planar graphs always have a vertex of degree at most $\lfloor 3.5\sqrt{k} \rfloor$ and consequently that every outer k-planar graph with n vertices has at most $\lfloor 3.5\sqrt{k} \rfloor n$ edges. This also means that the vertices of every outer k-planar graph can always be colored with $\lfloor 3.5\sqrt{k} \rfloor + 1$ colors so that no two endpoints of the same edge have the same color. Such problem is known as the *coloring* problem and it is one of the fundamental problems in graph theory. To complement our upper bound we show that an outer k-planar complete graph can have at most $(\lfloor \sqrt{4k+1} \rfloor + 2)$ vertices, therefore, $(\lfloor \sqrt{4k+1} \rfloor + 2)$ colors is necessary to color the complete graph.

We show further that each outer k-planar graph has a *balanced vertex separator* of size at most $2k + 3$, that is, it has a subset of vertices of size at most $2k + 3$ such that after removing this subset the graph falls apart into components of roughly equal sizes; we define this notion more formally in Section 2.1. We further show that for each fixed k, we can test outer k-planarity in quasi-polynomial time. Our recognition algorithm uses the fact that each outer k-planar graph has a balanced separator of size at most $2k + 3$. According to the *Exponential Time Hypothesis (ETH)*, no quasi-polynomial algorithm exists for a problem which is NP-hard, therefore, our algorithm implies that testing outer k-planarity is not NP-hard.

We compare the class of outer k-quasi-planar graphs to other graph classes, in particular, to the class of planar graphs. In addition, we observe simple bounds on the *page number* of outer k-quasi-planar graphs, that is, the minimum number of half-planes, called *pages*, to draw the graph planarly if all the vertices are on a line where all the pages intersect.

Finally, we restrict outer k-planar and outer k-quasi-planar drawings to *full* drawings (where no crossing appears on the boundary), and to *closed* drawings (where the vertex sequence

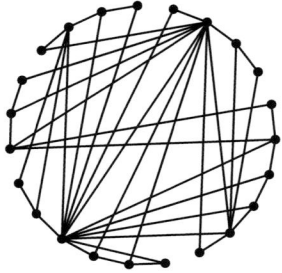 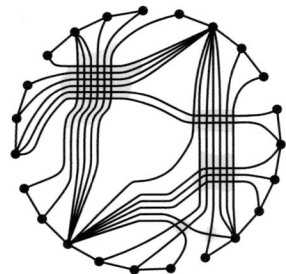

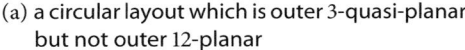

(a) a circular layout which is outer 3-quasi-planar but not outer 12-planar

(b) a bundled drawing of the drawing in Figure 1.4a; each crossing occurs between two bundles, that is, as a bundled crossing

Figure 1.4: A circular layout and its bundled drawing.

on the boundary is a cycle in the graph). For each k, we express *closed outer k-planarity* and *closed outer k-quasi-planarity* in *extended monadic second-order logic*; see Section 2.4 for the definition. Thus, since outer k-planar graphs have bounded treewidth (many problems on graphs with bounded treewidth are efficiently solvable, for the definition of the parameter see Section 2.1), closed outer k-planarity is linear-time testable by Courcelle's Theorem (see Section 2.4). We leverage this result to further show that full outer k-planarity can also be tested in linear time.

This chapter is based on joint work with Steven Chaplick, Giuseppe Liotta, Andre Löffler, and Alexander Wolff [CKL⁺18].

Bundled Crossing Minimization in Circular Layouts

Sometimes a graph may require a lot of crossings to be drawn, and no matter how we optimize the edge crossings (minimize the number of crossings or optimize the crossing angle), the drawing may still look cluttered and messy. With this in mind, in Chapter 7 we consider an effective way to reduce clutter in a drawing that has (many) crossings, by grouping edges that travel in parallel into *bundles*. Each edge can participate in many such bundles. Any crossing in this bundled drawing occurs between two bundles, i.e., as a *bundled crossing*; see for example Figure 1.4b. In this context we consider the problem of bundled crossing minimization: A graph is given and the goal is to find a bundled drawing with at most k bundled crossings.

We show that the problem is NP-hard when we require a *simple* drawing (that is, edges are not allowed to self-intersect and any two edges are not allowed to intersect twice), settling an open question by Fink et al. [FHSV16]. Our main result is an algorithm that, given a graph G and a natural number k, computes a simple circular layout with k bundled crossings if one exists. This algorithm is *fixed parameter tractable (FPT)* in k, that is, it has runtime $O(f(k)n^c)$, where n is the size of the graph, c is a constant, and $f(\cdot)$ is a computable function which only depends on k; see Section 2.3 for a formal definition. This answers an open question by Alam et al. [AFP16].

These results make use of the connection between bundled crossings and graph *genus* (the parameter that characterizes the surface that the graph needs to be drawn without crossings; for a formal definition see Section 2.1) and, as well as Chapter 6, extended monadic second-order logic (see Section 2.4).

We also consider bundling crossings in a given drawing, in particular, for *storyline visualizations*, that is, a set of x-monotone curves where each curve cannot self-intersect, but a pair of curves is allowed to intersect each other multiple times. The storyline literature considers the number of characters m to be small and the number of crossings to be large. We show that computing the *bundled crossing number* (that is, the minimum number of bundled crossings in the drawing) of a given storyline visualization is fixed parameter tractable in m.

This chapter is based on joint work with Steven Chaplick, Thomas C. van Dijk, Ji-won Park, Alexander Ravsky, and Alexander Wolff [CvDK$^+$19].

Preliminaries

This chapter provides some basic terminology, definitions, and concepts in graph theory, graph drawing, and algorithms used in this book. For a comprehensive introduction into these topics we refer to books designated to this matter.

2.1 Graphs

In this section we initially define graph theory concepts which are common in most graph theory areas and then we define graph parameters which are more specific but still relevant to several chapters of this book. A basic introduction to graph theory can be found in the book *Introduction to Graph Theory* by Trudeau [Tru93]. For introduction to graph algorithms we refer the reader to *Algorithmic Graph Theory* by Gibbons [Gib85].

A *graph* G is defined as a pair (V, E) where V is a non-empty set of vertices and $E \subseteq V \times V = \{(u, v) \subseteq V \mid u \neq v\}$ is a non-empty set of edges. Here we consider (u, v) to be the same as (v, u). For a graph G we denote its set of vertices as $V(G)$ and its set of edges as $E(G)$. A *multi-graph* is a graph G where there can be multiple edges between the same two vertices or edges between one vertex, called *loops*, that is, edges of type (v, v), $v \in V(G)$. The set of edges $E(G) \subseteq V(G) \times V(G) = \{(u, v) \subseteq V(G)\}$ of a multi-graph is a multi-set, that is, its elements can repeat. A graph which is not a multi-graph is sometimes also called a *simple* graph. In this book we mostly deal with simple graphs, therefore, whenever we say a graph we mean a simple graph, unless stated otherwise. For an edge $e = (u, v)$ we call u and v the *endpoints* of e. A graph with n vertices can contain at most $\binom{n}{2} = n(n-1)/2$ edges. Two vertices u and v of a graph G such that $(u, v) \in E(G)$ are called *adjacent* or *neighbors*. For an edge $e = (u, v)$ we say that the vertices u and v are incident to e. For a vertex v of the graph G, the *degree* $\deg(v) := |\{u \in V \mid (u, v) \in E(G)\}|$ is the number of vertices adjacent to v.

A *graph* G is said to be *directed* if the set of edges $E(G) \subseteq V(G) \times V(G) = \{(u, v) \subseteq V(G)\}$ consists of ordered pairs of vertices. In this case we distinguish between (u, v) and (v, u). We call an edge $e = (u, v)$ *outgoing* for vertex u and *incoming* for vertex v. In this book we mostly deal with *undirected* graphs (that is, graphs that are not directed), thus, whenever we say a graph we mean an undirected graph unless stated otherwise.

We call G' a *subgraph* of a graph G if $V(G') \subseteq V(G)$ and $E(G') \subseteq E(G)$. The subgraph G' is *induced* by $V(G')$ if $E(G')$ contains exactly the edges of G whose endpoints are both in $V(G')$. The induced graph is denoted by $G[V']$ for $V' = V(G')$.

A *subdivision* of a graph G is a graph G' obtained by replacing every edge by a path of some length. We say that we *contract* an edge (u, v) if we replace it by a single vertex w that is adjacent to the neighbors of u and v.

A *path* is a graph P with $V(P) = \{v_0, v_1, \ldots, v_k\}$, $k > 0$, and $E(P) = \{(v_i, v_{i+1}) \mid 0 \leq i \leq k-1\}$ and P is called a *cycle* if $v_0 = v_k$, k is called the *length* of P. The path P is called *simple* if all vertices are pairwise different (with the exception of $v_0 = v_k$, if P is a cycle).

A *Hamiltonian path* in a graph G is a path visiting every vertex of G exactly once. Similarly a *Hamiltonian cycle* in a graph G is a cycle visiting every vertex of G exactly once. A graph is called *Hamiltonian* if it has a Hamiltonian cycle.

A graph G is called *connected* if for every two vertices $u, v \in V(G)$ there exists a path from u to v in G; otherwise, G is called *disconnected*. A *connected component* of G is a connected subgraph $G[V']$ induced by a maximal subset $V' \subseteq V(G)$, that is, no other vertex u in $V(G) \setminus V'$ can be added to V' so that $G[V' \cup \{u\}]$ is connected.

A graph G is called *k-vertex-connected* if after the removal of any $k - 1$ vertices it still is connected. Similarly G is *k-edge-connected* if after the removal of any $k - 1$ edges it still is connected. If we do not specify whether a graph is vertex or edge connected and simply say that a graph is *k-connected* we mean that it is k-vertex-connected. A *cutvertex* of a graph G is a vertex v of G such that after its removal the graph is disconnected. We say that a graph G is *biconnected* or *triconnected* if it is 2-connected or 3-connected respectively.

In the following we list some of the most classic graphs in graph theory.

A graph with n vertices and with $n(n - 1)/2$ edges is called a *complete* graph and denoted as K_n. On the other end a connected graph with no cycle is called a *tree* and it contains exactly $n - 1$ edges. A graph which is a collection of trees is called a *forest*. Each tree has a unique simple path between every pair of vertices. For a tree T a vertex $v \in V(T)$ is called a *leaf* if $\deg(v) = 1$. A *caterpillar* is a tree that consists of a path, called *spine*, and leaves that are connected to the spine. A path (that we defined above) can be regarded as a special type of a tree or a caterpillar.

A graph G is called *bipartite* if the set of vertices consists of two disjoint sets A and B with $V(G) = A \cup B$ and $A \cap B = \varnothing$ and each edge is incident to a vertex of each set, that is, $E(G) = \{(u, v) \subseteq V(G) \mid u \in A, v \in B\}$. A bipartite graph with maximum degree 1 is called a *matching* and a bipartite graph with the maximum number, that is, nm of edges where $n = |A|$ and $m = |B|$ is called a *complete bipartite* graph and denoted as $K_{n,m}$.

Coloring and degeneracy. One of the most fundamental problems in graph theory is the *coloring problem*, where a graph G is given and we have to find the minimum number of colors to color the vertices of G so that no edge of G is incident to vertices of the same color. Such number is called the *chromatic number* of G. Providing bounds on the chromatic number of graphs from a certain graph class is one of the classical ways to characterize the graph class.

Graphs in which every subgraph has a vertex of degree at most d can be inductively $d + 1$ colored by simply removing a vertex of degree at most d. Thus, this property is of particular interest when dealing with coloring problems and such graphs are called *d-degenerate* [LW70]. Note that *outerplanar* graphs are 2-degenerate, and *planar* graphs are 5-degenerate; for definitions of planar and outerplanar graphs see Section 2.2.

Treewidth. A parameter *treewidth*, well known in algorithmic graph theory, was introduced by Bertelé and Brioschi [BB72] and then later rediscovered (and popularized) by Robertson and Seymour [RS84]. A *tree decomposition* of a graph G is a pair (X, T), where T is a tree and $X = \{X_i \mid i \in V(T)\}$ is a family of subsets of $V(G)$, called *bags*, such that (1) for all $v \in V(G)$, the set of nodes $T_v = \{i \in V(T) \mid v \in X_i\}$ induces a non-empty connected subtree of T, and (2) for each edge $uv \in E(G)$ there exists $i \in V(T)$ such that both u and v are in X_i. The maximum of $|X_i| - 1$, $i \in V(T)$, is called the *width* of the tree decomposition. The *treewidth*, $\text{tw}(G)$, of a graph G is the minimum width over all tree decompositions of G.

Graphs with bounded treewidth are of special interest because many problems on such graphs can be solved efficiently. In particular, for such graphs we can test efficiently graph properties that have a compact formulation in MSO_2 logic (see the definition of MSO_2 logic and Theorem 2.1 in Section 2.4).

Separators. Another useful concept that we utilize in several chapters is that of a *vertex separator*. Unfortunately, the precise definition of the concept is inconsistent throughout the literature. An attempt to unify the existing definitions was done by Harvey and Wood [HW17]. We use their notation. We will consider two different types of vertex separators; one in Chapter 3 and the other one in Chapter 6.

For a graph G, a set $W \subseteq V(G)$, and $c \in [\frac{1}{2}, 1)$, a $(k, W, c)^*$-*separator* is a subset of vertices $S \subset V(G)$, with $|S| \leq k$, such that each connected component of $G - S$ contains at most $c|W|$ vertices of W. The number $|S|$ is the *size* of the separator.

For the purpose of Chapter 3 we need a type of the $(k, W, c)^*$-*separator* as used by Flum and Grohe [FG06]. If $c = \frac{1}{2}$, then a $(k, W, c)^*$-separator S is called a *balanced W-separator*. The *W-separation number* $\text{sep}_W(G)$ is the minimum k such that G has a balanced W-separator S with $|S| = k$. This separation number is related to treewidth. According to [FG06, Theorem 11.17], $\text{sep}_W(G) \leq \text{tw}(G) + 1$ for any $W \subseteq V(G)$ and, on the other hand, $\text{tw}(G) \leq 3k + 2$ if $\text{sep}_W(G) \leq k$ for every W with $|W| = 2k + 3$.

In Chapter 6 we define the separation number similarly to that defined by Fox [Fox11]. For a given graph G the *separation number* $\text{sn}(G)$ is the minimum integer k such that, for each subgraph H of G, there exists a $(k, V(H), \frac{2}{3})^*$-separator for H. Such a separator is called a *balanced separator*; note that, unlike for balanced W-separators, the factor here is $\frac{2}{3}$ instead of $\frac{1}{2}$. For a graph G the separation number $\text{sn}(G)$ is also related to the treewidth $\text{tw}(G)$ of G, namely, $\text{sn}(G) \leq \text{tw}(G) + 1$ [HW17] and, as Dvořák and Norin [DN14] recently showed, $\text{tw}(G) \leq 105\text{sn}(G)$.

2.2 Graph Drawing

In this section we define some of the Graph Drawing concepts. For a reference we suggest the books *Graph Drawing: Algorithms for the Visualization of Graphs* by Di Battista et al. [DETT99], *Drawing Graphs: Methods and Models* edited by Kaufmann and Wagner [KW01], *Planar Graph Drawing* by Nishizeki and Rahman [NR04], and the *Handbook of Graph Drawing and Visualization* edited by Tamassia [Tam13].

A *drawing* D of a graph G is a mapping that maps each vertex $v \in V(G)$ to a point in some surface and each edge $uv \in E(G)$ to a simple open Jordan curve on the surface such that the endpoints of this curve are $D(u)$ and $D(v)$. For convenience, our notation will not distinguish between the entities (vertices and edges) of an abstract graph and the geometric objects (points and curves) representing them in a drawing. A *crossing* of two edges is the common point of their interior. A shared endpoint of two edges is not considered a crossing. In general, we assume that no three edges cross in a single point; edges do not pass through vertices; if two edges share a point, they must cross at this point, that is, they cannot *touch* at this point; and there is only a finite number of crossing points in the drawing. If any two edges share at most one point in their interior and no edge is allowed to self-intersect, then the drawing is called *simple*. Since in this book we mostly deal with simple drawings whenever we say a drawing, we mean a simple drawing unless stated otherwise.

If a graph G can be drawn without crossings it is called *planar*. The *planarization* G' of a drawing D with crossings is a planar graph obtained from D by creating a vertex at each crossing point and connecting two such vertices or vertices of the graph if they are consecutive along an edge in D.

The drawing D in a surface $\mathfrak{S}$ subdivides this surface into topologically connected regions of $\mathfrak{S} \setminus D$. These regions are called *faces* of D. If $\mathfrak{S}$ is unbounded, the unbounded face of D is called the *outer* face. A vertex or an edge is called *incident* to a face f if it lies on the boundary of f. A planar graph where all vertices are incident to the outer face is called *outerplanar*.

A planar graph can have many different drawings but some of them might have a lot of common features. To capture the similarity between drawings the notion of *embedding* is introduced. An embedding of a planar graph is a *rotation system*, that is, the circular order of the incident edges around a vertex in some drawing of this graph together with a specified outer face in this embedding. Drawings with the same embedding are called *equivalent*. Equivalent drawings share some essential properties, for example, they have the same set of faces. An *embedding* of a graph which is not necessarily planar is an equivalence class of drawings whose planarizations have the same planar embedding.

To specify a drawing of a graph beyond the embedding the notions of the *drawing style* or *layout* are introduced which determine how the edges or vertices are drawn. The most common style is to draw edges as *straight-line* segments. One of the generalizations of straight-line drawings is *poly-line drawings* where an edge is drawn as a finite sequence of straight-line segments touching at endpoints, these endpoints are called *bends*. Another generalization is *circular-arc drawings* where each edge is drawn as a circular arc, a straight-line edge is considered an arc of infinite radius.

Once a drawing style is fixed we can evaluate qualitative properties of a drawing which in turn allows us to compare different drawings in the same style. Such qualitative properties of a drawing are called *quality measures*. A *quality measure* of a graph is the best value of the measure over all drawings of the graph in some fixed style. For example, a classic measure is the *drawing area*. If the vertices of a graph are drawn on a grid, the area is determined by the *width* and the *height* of the grid. If the vertices in the drawing are not necessarily on a grid, the area can be determined by the ratio between the maximum distance among all pairs of vertices to the smallest distance among all pairs of vertices. Drawings with small area are preferable as they can fit into a small drawing canvas or screen. Another common quality measure is *angular resolution* of a drawing which is the smallest angle formed by any pair of edges incident to the same vertex. Drawings with large angular resolution are preferable as then it is easier to distinguish incident edges from each other. A drawing has *perfect angular resolution* if the edges are equally spaced around each vertex.

Recently, a new type of quality measure was introduced: the number of geometric objects that are needed to draw a graph given a certain style. This measure is also known as the the *visual complexity* [Sch15] of a drawing. Note that several edges of a graph can be drawn on the same geometric object. Drawings of large visual complexity tend to be difficult to perceive visually [KMS18], thus the goal is to minimize visual complexity. Two classical examples of quality measures regarding visual complexity are the *segment number* [DESW07], that is, the minimum number of straight-line segments over all straight-line drawings of a graph and the *arc number* [Sch15], that is, the minimum number of circular arcs over all circular-arc drawings of a graph.

Since the fundamental topic of Beyond Planarity is crossing optimization a lot of quality measures deal with crossings. The most direct way to tackle crossings is to avoid them completely. This is however impossible in the plane for a non-planar graph. Thus for non-planar graphs we need either to allow multiple surfaces to contain the drawing or to modify the surface of the drawing.

One of the parameters that is related to the first approach is called the *thickness* of a graph, which, for a given graph G, is the smallest number of planar graphs whose union is G. Given decomposition of G into k planar subgraphs we can draw it on k surfaces so that each planar subgraph is in a different surface and these surfaces intersect only in the vertices of the graph that are, for example, in the same plane. A similar but more specific parameter is the smallest number of outerplanar graphs whose union is G. This parameter is called the *book thickness* of a graph. The name comes from an alternative representation, called *book embedding*, where the vertices of G lie on some common line, called the *spine* of the book, and the edges of G are drawn in some halfplanes that all intersect at the spine, called the *pages* of the book. Book thickness is also sometimes called *page number*. It is known that planar graphs have page number 4 [Yan89] and there are planar graphs that require 4 pages [Yan20, BKK$^+$20]. Note that these parameters are similar to visual complexity, in that, they record the smallest number of surfaces (the objects regarded by visual complexity) containing the drawing.

Further if we specify the type of edges that we use in the drawing and the type of surfaces containing a drawing we obtain different quality measures. The *affine cover number* [CFL$^+$16] is the minimum number of l-dimensional affine subspaces that together cover a crossing-free straight-line drawing of the given graph in d-dimensional space, for integers l and d such that $0 < l < d$. Note that if $d = 2$ and $l = 1$, the affine cover number is defined for planar graphs only and it is similar to the segment number, with the difference that the attention is on the number of lines containing the drawing rather than segments (several different segments can be contained in the same line). It is also worth mentioning that for $d = 2$ and $l = 1$ the optimal affine covers are related to *minimum-line drawings* [DMNW11], that is, minimum-segment drawings whose edges lie in the union of the smallest number of straight lines (among all minimum-segment drawings). However, the minimum-line drawings are essentially different from ρ_2^1-optimal drawings since there are graphs that do not have a ρ_2^1-optimal cover with the minimum number of segments; see Section 3.6.

Similarly to the affine cover number the *spherical cover number* of a graph G is the minimum number of l-dimensional spheres in $\mathbb{R}^d$ such that G has a crossing-free circular-arc drawing that is contained in the union of these spheres, for integers l and d such that $0 < l < d$. Again if $d = 2$ and $l = 1$, the spherical cover number is defined for planar graphs only and is similar to the arc number.

In contrast to allowing multiple planes or spheres in order to draw a non-planar graph without crossings another approach is to modify the surface on which the graph is drawn. This can be done by adding handles to the surface. The number of handles on a surface is called the *genus* of the surface. Alternatively genus of a surface can be defined as an integer representing the maximum number of cuttings along non-intersecting closed simple curves without rendering the resulting surface disconnected. A *genus* of a graph is the minimum number of handles on a surface where the graph can be drawn without crossings. This parameter has been known for much longer than the affine and spherical cover numbers [FMR79, Tho89].

If a graph admits a drawing with only few crossings, then there is hope to draw this graph with crossings so that these crossings do not affect the readability of the drawing too much. Thus, quality measures regarding crossings are considered. Restrictions on such quality measures also define different graph classes, that is, classes of graphs that can be drawn so that the crossings are restricted in some way.

For example, graphs that can be drawn with at most k crossings per edge are called *k-planar graphs*. Graphs that can be drawn with no k pairwise crossing edges are called *k-quasi-planar graphs*. Further these restrictions can be combined with a layout. If all the vertices in a drawing are in a convex position, for example on a circle, and the edges are drawn inside the disk of the circle, then the drawing is called a *circular layout*. Such drawings are also known as *convex drawings*. If we in addition insist that the drawing is planar, then the drawing (and the corresponding graph) is outerplanar. If instead we require that the drawing is k-planar or k-quasi-planar, then it (and the corresponding graph) is called *outer k-planar* or *outer k-quasi-planar* respectively.

Another common quality measure regarding crossings is called *crossing angle resolution*, which is the smallest angle formed by any pair of crossing edges at any of their crossings. Again, restrictions on a measure yield different graph classes. For example, graphs that have straight-line drawings with crossing angle resolution bounded from below by some constant or with *right-angle crossings* only. The latter are known as RAC graphs. Requirements on a measure can also be combined with other different drawing styles obtaining classes of graphs that can be drawn with poly-line edges and right-angle crossings or circular-arc edges and right-angle crossings.

A different approach to handle crossings was introduced by Holten [Hol06] where edges that travel in parallel in a drawing with (many) crossings are grouped into *bundles*. A crossing of two bundles can involve a lot of edge crossings but we count it as one *bundled crossing*. For certain drawings the number of bundled crossings can be significantly smaller than the number of edge crossings. Interestingly bundled crossings are related to genus, as we can resolve a crossing of two bundles by routing the edges of one bundle via a handle going over the other bundle in the surface of the drawing. We discuss this relation in more detail in Chapter 7.

2.3 Complexity

In this section we provide some of the the the basics of algorithm complexity and parameterized complexity theories used in this book. For an introduction to algorithm theory we refer to *Introduction to Algorithms* by Cormen et al. [CLRS09]. An introduction to complexity theory can be found in *Computers and Intractability: A Guide to the Theory of NP-Completeness* by Garey and Johnson [GJ79] and for an introduction to parameterized complexity theory we refer to the book *Invitation to Fixed-Parameter Algorithms* by Niedermeier [Nie06].

Runtime of an algorithm is the amount of resources required to execute it on an instance of certain size. It is measured in the number of steps that the algorithm takes to solve the instance. When analyzing runtime of an algorithm, we usually analyze the *asymptotic runtime*, which is described by the *big O notation*. Let $f : \mathbb{N} \to \mathbb{N}$ be a function that maps the input size of a problem to the time that the algorithm needs to solve the problem. Then the class of functions that asymptotically grow at most as fast as f is denoted by

$$O(f) = \{g : \mathbb{N} \to \mathbb{N} \mid \exists c > 0, n_0 \in \mathbb{N} \quad \forall n \geq n_0 : g(n) \leq cf(n)\}.$$

We say that the runtime g of an algorithm is in $O(f)$, if $g \in O(f)$. If the runtime of an algorithm is in $O(n^c)$ where c is some constant, then the algorithm is called a *polynomial-time* or *efficient* algorithm. If the runtime is of the form $2^{\text{polylog}(n)}$ then the corresponding algorithm is called *quasi-polynomial*.

The class of problems for which there exists a deterministic algorithm that solves the problem in polynomial time is denoted as P. Similarly the class of problems for which there exists a nondeterministic algorithm that solves the problem in polynomial time is denoted as NP. It holds that $P \subseteq NP$, however, it is not known whether $P = NP$ or $P \subsetneq NP$.

A problem A is called NP-*hard* if every problem $B \in NP$ can be transformed to A in polynomial time. Such a transformation from B to A mapping any instance X of B to an instance Y of A so that there is a valid solution for X if and only if there is a valid solution for Y is called *reduction*. Therefore, if at least one problem in NP can be solved efficiently, then we can solve every other problem efficiently. The current conjecture, though, is that $P \neq NP$, and thus, we do not expect that an NP-hard problem can be solved efficiently. A problem is called NP-*complete* if it is NP-hard and in NP.

Usually, NP-hardness is proved by reduction from a known NP-hard problem. In the following we list some of the known NP-hard problems used in this book.

3SAT (3-Satisfiability)	
Given:	a set of Boolean variables U and a set C of clauses such that each clause contains three literals from U.
Find:	a truth assignment to the variables so that each clause is satisfied.

The problem 3SAT is one of the "core", as listed by Garey and Johnson [GJ79], NP-complete problems most frequently used for reduction. It has a lot of variants that are particularly applicable in Graph Drawing. Below we state one of these variants.

NAE3SAT (Not-All-Equal-3-Satisfiability)	
Given:	a set of Boolean variables U and a set C of clauses such that each clause contains three literals from U.
Find:	a truth assignment to the variables so that each clause contains at least one true literal and at least one false literal.

The problem NAE3SAT can be mechanically simulated by a paradigm called the "Logic Engine" [DETT99, Section 11.2]. This paradigm is often used to prove complexity of Graph Drawing problems.

GENUS (Determining graph genus)	
Given:	a graph G.
Find:	the genus of the graph G.

The problem GENUS is known to be NP-hard even though if the genus is fixed, there exists an algorithm with linear runtime with respect to the size of the input [Tho89, Moh99, KMR08].

As a way to deal with NP-hard problems there are several complexity classes defined for problems that can be solved in time depending not only on the size n of the instance but also on some parameter of the instance k so that the runtime of the algorithm is dominated by the parameter k and is polinomial in n if k is fixed.

A problem A is called *fixed parameter tractable* or *FPT* if there is an algorithm that correctly decides, for input X and the parameter k, whether X is a solution to A in time $O(f(k)n^c)$, where n is the size of the input, that is, $n = |X|$, c is a constant, and $f(\cdot)$ is a computable function which only depends on k. The algorithm with such a runtime is called an *FPT* algorithm.

A problem A is in the class XP if there is an algorithm that correctly decides, for input X and the parameter k, whether X is a solution to A in time $O(n^{f(k)})$, where n is the size of the main part of the input, that is, $n = |X|$ and $f(\cdot)$ is a computable function which only depends on k. The algorithm with such a runtime is called an XP algorithm.

For some problems, however, it is believed that algorithms with exponential runtime are unavoidable. The *Exponential Time Hypothesis (ETH)* [IP01] is a complexity theoretic assumption defined as follows. For $k \geq 3$, let $s_k = \inf\{\delta : \text{there is an } O(2^{\delta n})\text{-time algorithm}$ to solve k-SAT$\}$. ETH states that for $k \geq 3$, $s_k > 0$, e.g., there is no quasi-polynomial time algorithm that solves 3SAT. So, finding a problem that can be solved in quasi-polynomial time and is also NP-hard, would contradict the ETH. In recent years, the ETH has become a standard assumption from which many conditional lower bounds have been proven [CFK$^+$15].

2.4 Monadic Second-Order Logic

In Chapters 6 and 7 we design several algorithms based on extended monadic second-order logic (MSO$_2$). In this section we introduce some basic definitions of the concept. For background on monadic second-order logic, we refer to the textbook of Courcelle and Engelfriet [CE12].

The class of formulas expressible in MSO$_2$ is defined as follows. *Extended monadic second-order logic* (MSO$_2$) is a subset of second-order logic that can be used to express certain graph properties. It is built from the following primitives:

- variables for vertices, edges, sets of vertices, and sets of edges;

- binary relations for: equality ($=$), membership in a set ($\in$), subset of a set ($\subseteq$), and edge–vertex incidence (I);

- standard propositional logic operators: $\neg$, $\wedge$, $\vee$, $\rightarrow$, and $\leftrightarrow$;

- standard quantifiers ($\forall$, $\exists$) which can be applied to all types of variables.

Note that MSO$_2$ differs from full second order logic in that it does not allow quantification over sets of sets, such as sets of pairs of vertices that are not subsets of the given edges. If we drop the "$_2$" then we have *monadic second-order logic* (MSO) where the only difference is that we are now not allowed to quantify over edge sets. Additionally, the convention for MSO formulas is to use a binary *adjacency function (adj$_G$)* instead of the incidence function as in the definition of MSO$_2$ above.

For a graph G and an MSO$_2$ (or MSO) formula ψ, we use $G \vDash \psi$ to indicate that ψ can be satisfied by G in the obvious way.

Another tool that will be useful for us in Chapter 7 is an *L-transduction*, which is the operation of constructing the model of one graph/structure from the model of another graph/structure in the language of the logic L. A formal treatment on transductions is given in the book of Courcelle and Engelfriet [CE12, Section 1.7.1, and Definitions 7.6 and 7.25]. For example, an MSO-transduction that constructs a graph G' (modelled for MSO, using a vertex set $V(G')$ and adjacency function $adj_{G'}$) by adding a universal vertex x to a given graph G (modelled for MSO$_2$, using a vertex set $V(G)$, edge set $E(G)$, and incidence function I_G) can be written as follows:

$$V(G') := \{x\} \cup V(G)$$
$$adj_{G'}(u, v) := (u \neq v) \wedge$$
$$\Big(\big((\exists e \in E(G)) \, (I_G(e, v) \wedge I_G(e, u))\big) \vee (x = u) \vee (x = v)\Big).$$

Graph properties that can be expressed as an MSO$_2$ formula of bounded size can be efficiently tested using Courcelle's theorem, provided that the given graph has bounded treewidth.

Theorem 2.1 (Courcelle [Cou90, CE12]). *For any integer $t \geq 0$ and any MSO$_2$ formula ψ of length ℓ, an algorithm can be constructed which takes a graph G with n vertices, m edges, and treewidth at most t and decides in $O(f(t, \ell) \cdot (n + m))$ time whether $G \vDash \psi$ where the function f from this time bound is a computable function of t and ℓ.*

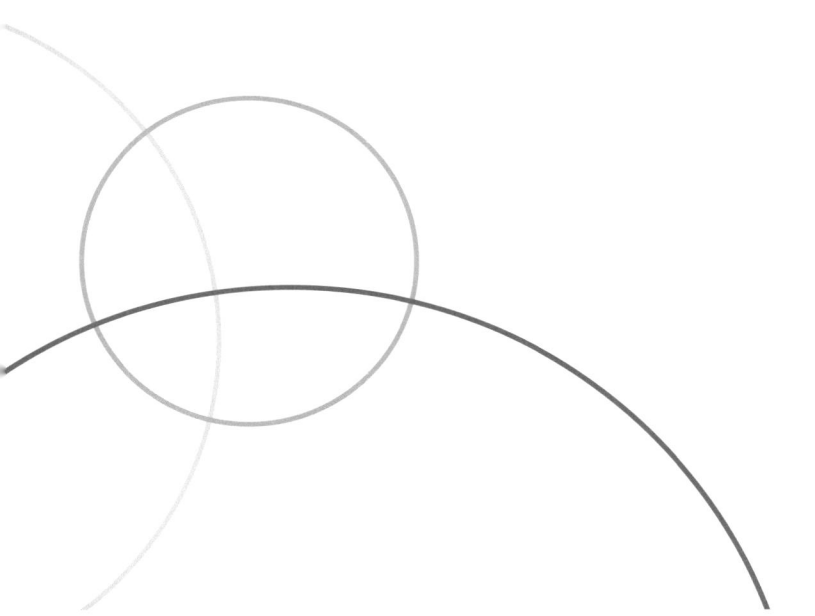

Part I

Optimizing Crossings
in Circular-Arc Drawings

Covering Graphs
with Few Circles and Few Spheres

A drawing of a given graph can be evaluated by many different quality measures depending on the concrete purpose of the drawing. Classical examples are the number of crossings, the ratio between the lengths of the shortest and the longest edge, or the angular resolution. Clearly, different layouts (and layout algorithms) optimize different measures. Hoffmann et al. [HvKKR14] studied ratios between optimal values of quality measures implied by different graph drawing styles. For example, there is a circular-arc drawing of the icosahedron with perfect angular resolution (that is, the edges are equiangularly spaced around each vertex), whereas the best straight-line drawing has an angular resolution of at most 15°, which yields a ratio of 72°/15° = 4.8. Hoffmann et al. also constructed a family of graphs whose straight-line drawings have unbounded edge–length ratio, whereas there are circular-arc drawings with edge–length ratios arbitrarily close to 3 [HvKKR14, Figures 4 and 6].

A few years ago, a new type of quality measure was introduced: the number of geometric objects that are needed to draw a graph given a certain style. Schulz [Sch15] coined this measure the *visual complexity* of a drawing. More concretely, Dujmović et al. [DESW07] defined the *segment number* $seg(G)$ of a graph G to be the minimum number of straight-line segments over all straight-line drawings of G. Similarly, Schulz [Sch15] defined the *arc number* $arc(G)$ with respect to circular-arc drawings of G and showed that circular-arc drawings are an improvement over straight-line drawings not only in terms of visual complexity but also in terms of area consumption; see Schulz [Sch15, Theorem 1]. Mondal et al. [MNBR13] showed how to minimize the number of segments in convex drawings of 3-connected planar graphs both on and off the grid. Igamberdiev et al. [IMS17] fixed a bug in the algorithm of Mondal et al. and compared the resulting algorithm to two other algorithms in terms of angular resolution, edge length, and face aspect ratio. Hültenschmidt et al. [HKMS18] studied the visual complexity of drawings of planar graphs. For example, they showed upper bounds for the number of segments and arcs in drawings of trees, triangulations, and general planar graphs. Recently, Kindermann et al. [KMS18] presented a user study showing that people without mathematical or computer science background prefer drawings that consist of few line segments, that is, drawings of low visual complexity. (Users with such a background had a slight tendency to prefer drawings that are more symmetric.) The study, however, was done for trees only.

Durocher et al. [DMNW11] investigated the complexity of computing minimum-segment drawings (and related problems). Among others, they showed that it is NP-hard to compute the segment number of plane graphs (that is, planar graphs with fixed embedding), even if the graphs have maximum degree 4. As an open problem, the authors suggested to study *minimum-line drawings*, which they define to be minimum-segment drawings whose edges lie in the union of the smallest number of straight lines (among all minimum-segment drawings).

Chaplick et al. [CFL⁺16] defined a similar quality measure, which they call the *affine cover number*. Given a graph G and two integers l and d with $0 < l < d$, they defined $\rho_d^l(G)$ to be the minimum number of l-dimensional affine subspaces that together cover a crossing-free straight-line drawing of G in d-dimensional space. It turned out that it suffices to consider $l \leq 2$ because otherwise $\rho_d^l(G) = 1$. In [CFL$^+$16] the authors also show that every graph can be drawn in 3-space as effectively as in high dimensional spaces, i.e., for any integers $1 \leq l \leq 3 \leq d$ and for any graph G, it holds that $\rho_d^l(G) = \rho_3^l(G)$. Note that, in general, (as already mentioned in Chapter 2) the minimum-line drawings mentioned above are different from ρ_2^1-optimal drawings since there are graphs that do not have a ρ_2^1-optimal cover with the minimum number of segments; see Example 3.1 in Section 3.6.

Among others, Chaplick et al. showed that the affine cover number can be asymptotically smaller than the segment number, constructing an infinite family of triangulations $(T_n)_{n>1}$ such that T_n has n vertices and $\rho_2^1(T_n) = O(\sqrt{n})$, but $\text{seg}(T_n) = \Omega(n)$. On the other hand, they showed that $\text{seg}(G) = O(\rho_2^1(G)^2)$ for any connected planar graph G. In a companion paper [CFL$^+$17], Chaplick et al. show that most variants of the affine cover number are NP-hard to compute.

Our contribution.   Combining the approaches of Schulz and Chaplick et al., we introduce the *spherical cover number* $\sigma_d^l(G)$ of a graph G to be the minimum number of l-dimensional spheres in $\mathbb{R}^d$ such that G has a crossing-free circular-arc drawing that is contained in the union of these spheres. Note that $\sigma_2^1(G)$ is defined for planar graphs only.

Firstly, we provide some basic observations and preliminary results that our work heavily relies on.

We obtain bounds for the spherical cover number σ_3^2 of the complete and complete bipartite graphs which show that spherical covers can be asymptotically smaller than affine covers; see Table 3.1 and Section 3.2.

Then we turn to platonic graphs, that is, to 1-skeletons of platonic solids; see Section 3.3. These graphs possess several nice properties: they are regular, planar and Hamiltonian. We use them as indicators to compare the above-mentioned measures of visual complexity; we provide bounds for their segment and arc numbers (see Table 3.2) as well as for their affine and spherical cover numbers (see Table 3.3). For the upper bounds, we present straight-line drawings with (near-) optimal affine cover number ρ_2^1 and circular-arc drawings with optimal spherical cover number σ_2^1; see Figures 3.4–3.6. We note that sometimes optimal spherical covers are more symmetric than optimal affine covers. For example, it seems that there is no symmetric drawing of the cube that is ρ_2^1-optimal, whereas there are symmetric σ_2^1-optimal drawings; see Fig. 3.4.

For general graphs, we present lower bounds for the spherical cover numbers by means of some combinatorial graph characteristics, in particular, by treewidth, balanced W-separator size, linear arboricity, and bisection width; see Section 3.4.

We decided to start with our more concrete (and partially stronger) results and postpone the structural observations to Section 3.4, although this means that we'll sometimes have to use forward references to Theorem 3.3, our main result in Section 3.4. Finally, we formulate a mixed-integer linear program (MIP) that yields lower bounds for the segment number of embedded planar graphs; see Section 3.5. For the platonic solids, the lower bounds (see Table 3.4) that we computed using the MIP turned out to be tight. We conclude with a few open problems.

3.1 Preliminary Results

In this section we state some preliminary results. Firstly, we note that any drawing with straight-line segments and circular arcs can be transformed into a drawing that uses circular arcs only.

Proposition 3.1. *Given a graph G and a drawing Γ of G that represents edges as straight-line segments or circular arcs on r l-dimensional planes or spheres in $\mathbb{R}^d$, there is a circular-arc drawing Γ' of G on r l-dimensional spheres in $\mathbb{R}^d$. In particular, $\sigma_d^l(G) \leq \rho_d^l(G)$ for any graph G and $1 \leq l < d$.*

Proof: Take an arbitrary sphere $S \subset \mathbb{R}^d$ that does not intersect any of the r spheres or planes that support the given drawing Γ of G. Without loss of generality, assume that S is centered at the origin. This implies that none of the spheres supporting Γ goes through the origin. Let ρ be the radius of S. Invert the drawing with respect to S by the map $x \mapsto \rho x / \|x\|$. (For a more formal definition of inversion see Chapter 4.) The resulting drawing is a circular-arc drawing of G on r l-dimensional spheres in $\mathbb{R}^d$. Indeed, using basic properties of the inversion (see, for instance, [Ede06] or [BEG11, Chapter 5.1]), it can be proved that this inversion transforms planes into spheres of the same dimension and preserves spheres, in other words, the set of images of points on a sphere forms another sphere of the same dimension. $\square$

Therefore, we may consider any line a "circle of infinite radius", any plane a "sphere of infinite radius", and any affine cover a spherical cover. By "line" we always mean a straight line.

Trivial bounds on $\sigma_3^1(G)$ follow from the fact that every circle is contained in a plane and that we have more flexibility when drawing in 3D than in 2D. Note again that $\sigma_2^1(G)$ and $\sigma_3^1(G)$ are only defined when G is planar.

Proposition 3.2. *For any graph G, it holds that $\rho_3^2(G) \leq \sigma_3^1(G)$. If G is planar, we additionally have $\sigma_3^1(G) \leq \sigma_2^1(G)$.*

The spherical cover number $\sigma_3^2(G)$ can be considered a characteristic of a graph G that lies between its *thickness* $\theta(G)$, which is the smallest number of planar graphs whose union is G, and its *book thickness* $\mathrm{bt}(G)$, which is the minimum number of pages (halfplanes) needed to draw the edges of G when the vertices lie on the *spine* of the book (the line that bounds all halfplanes).

Proposition 3.3. *For every graph G, it holds that $\theta(G) \leq \sigma_3^2(G) \leq \lceil \mathrm{bt}(G)/2 \rceil$.*

Proof: Each sphere covers a planar subgraph of G, so $\sigma_3^2(G)$ is bounded from below by $\theta(G)$. On the other hand, given a book embedding of a graph G with the minimum number of pages (equal to $\mathrm{bt}(G)$), we put the vertices from the spine along a circle which is the common intersection of $\lceil \mathrm{bt}(G)/2 \rceil$ spheres; see Fig. 3.1a. Then, for each page, we draw all its edges as arcs onto a hemisphere. Thus, we obtain a drawing witnessing $\sigma_3^2(G) \leq \lceil \mathrm{bt}(G)/2 \rceil$. $\square$

To bound $\sigma_2^1(G)$ and $\sigma_3^1(G)$ for the platonic solids in Section 3.3 from below we use a combinatorial argument similar to that in Lemma 7(a) and Lemma 7(b) in [CFL+16] which is based on the fact that each vertex of degree at least 3 must be covered by at least two lines and two lines can cross at most once, therefore, providing a lower bound on the number of lines given the number of vertices. We use a similar argument together with the fact that two circles can cross at most twice.

Proposition 3.4. *For any integer* $d \geq 1$ *and any graph* G *with* n *vertices and* m *edges, the following bounds hold:*

(a) $\sigma_d^1(G) \geq \frac{1}{2}\left(1 + \sqrt{1 + 2\sum_{v \in V(G)} \left\lceil \frac{\deg v}{2}\right\rceil \left(\left\lceil \frac{\deg v}{2}\right\rceil - 1\right)}\right);$

(b) $\sigma_d^1(G) \geq \frac{1}{2}\left(1 + \sqrt{2m^2/n - 2m + 1}\right)$ *for any graph* G *with* $m \geq n \geq 1$.

3.2 Complete and Complete Bipartite Graphs

In this section we investigate the spherical cover numbers of complete graphs and complete bipartite graphs. We first cover these graphs by spheres then by circular arcs, in 3D (and higher dimensions).

Theorem 3.1.

(a) *For any* $n \geq 3$, *it holds that* $\lfloor (n+7)/6 \rfloor \leq \sigma_3^2(K_n) \leq \lceil n/4 \rceil$.

(b) *For any* $1 \leq p \leq q$, *it holds that* $pq/(2p + 2q - 4) \leq \sigma_3^2(K_{p,q}) \leq p$ *and, if additionally* $q > p(p-1)$, *it holds that* $\sigma_3^2(K_{p,q}) = \lceil p/2 \rceil$.

Proof: (a) By Proposition 3.3, $\theta(K_n) \leq \sigma_3^2(K_n) \leq \lceil \mathrm{bt}(K_n)/2 \rceil$. It remains to note that, e.g., Duncan [Dun11] showed that $\theta(K_n) \geq \lfloor (n+7)/6 \rfloor$ and Bernhart and Kainen [BK79] showed that $\mathrm{bt}(K_n) = \lceil n/2 \rceil$.

(b) Again, it suffices to bound the values of the graph's thickness and book thickness. It can be easily shown that $\mathrm{bt}(K_{p,q}) \leq \min\{p, q\}$. On the other hand, Harary [Har15, Section 7, Theorem 8] showed that $\theta(K_{p,q}) \geq pq/(2p + 2q - 4)$. Due to Proposition 3.3, $\theta(K_{p,q}) \leq \sigma_3^2(K_{p,q}) \leq \min\{p, q\} \leq p$. In particular, if $q > p(p-1)$ then $\mathrm{bt}(K_{p,q}) = p$, due to Bernhart and Kainen [BK79, Theorem. 3.5], and $\lceil pq/(2p + 2q - 4) \rceil = \lceil p/2 \rceil$, so in this case $\sigma_3^2(K_{p,q}) = \lceil p/2 \rceil$. $\square$

Theorem 3.1 implies that any n-vertex graph G has $\sigma_3^2(G) \leq \lceil n/4 \rceil$.

On the other hand, given a graph G, we can bound $\sigma_3^1(G)$ from below in terms of the *bisection width* $\mathrm{bw}(G)$ of G, that is, the minimum number of edges between the two sets (W_1, W_2) of a *bisection* of G that is, a partition of the vertex set $V(G)$ of G into two sets W_1 and W_2 with $|W_1| = \lceil n/2 \rceil$ and $|W_2| = \lfloor n/2 \rfloor$.

Proposition 3.5. *For any graph* G *and* $d \geq 2$, *it holds that* $\sigma_d^1(G) \geq \mathrm{bw}(G)/2$.

Proof: The proof is similar to the proof in Theorem 9(a) in [CFL$^+$16]. It is based on the fact that for any finite set of points in $\mathbb{R}^d$, there is a hyperplane that bisects the point set into two almost equal subsets (that is, one subset may have at most one point more than the other).

Given a drawing of G with σ arcs, a hyperplane bisecting $V(G)$ can cross at most 2σ edges since a hyperplane can cross an arc at most twice. $\qquad\square$

Next we analyze the bisection width of the complete (bipartite) graphs.

Proposition 3.6. *For any n, p, and q, $\mathrm{bw}(K_n) = \lfloor n^2/4 \rfloor$ and $\mathrm{bw}(K_{p,q}) = \lceil pq/2 \rceil$.*

Proof: Let (W, W') be a bisection of K_n such that $|W| = \lfloor n/2 \rfloor$. Then the width of this bisection is $\lfloor n^2/4 \rfloor$.

Now let $P \cup Q = V(K_{p,q})$ be the bipartition of $K_{p,q}$, and let (W, W') be a bisection of $K_{p,q}$ that contains r vertices from P and s vertices from Q (with $r + s = \lfloor (p+q)/2 \rfloor$). Then the width of this bisection is $r(q - s) + s(p - r)$. The minimum of this value can be found by a routine calculation of the minimum of a quadratic polynomial on the grid over the possible values of r and s. $\qquad\square$

Theorem 3.2. *For any positive integers n, p, and q, it holds that*

(a) $\lfloor n^2/8 \rfloor \leq \sigma_3^1(K_n) \leq (n^2 + 5n + 6)/6$ and

(b) $\lceil pq/4 \rceil \leq \sigma_3^1(K_{p,q}) \leq \lceil p/2 \rceil \lceil q/2 \rceil$.

Proof: The lower bounds follow from Proposition 3.5 and 3.6.

To show the upper bound for $\sigma_3^1(K_n)$, we use a partition of K_n into (mutually edge-disjoint) subgraphs of K_3 (that is, copies of K_3, paths of length 2, and single edges). Using Steiner triple systems, one can show that $(n^2 + 5n + 6)/6$ subgraphs suffice [CFL+16, Theorem 12]. For distinct points a, b, and c, let $L(a, b)$ be the line through a and b and let $C(a, b, c)$ be the (unique) circle through a, b, and c. For $n \leq 3$, it is clear how to draw K_n. For $n \geq 4$, we iteratively construct a set P of n points in $\mathbb{R}^3$ satisfying the following conditions:

- no four distinct points of P are coplanar,

- for any five distinct points $p_1, \ldots, p_5 \in P$, it holds that $C(p_1, p_2, p_5) \cap C(p_3, p_4, p_5) = \{p_5\}$ and $L(p_1, p_2) \cap C(p_3, p_4, p_5) = \varnothing$,

- for any six distinct points $p_1, \ldots, p_6 \in P$, it holds that $C(p_1, p_2, p_3) \cap C(p_4, p_5, p_6) = \varnothing$.

It can be checked that these conditions forbid only a so-called *nowhere dense set* of $\mathbb{R}^3$ to place the next point of P, so we can always continue. Finally, we map the vertices of K_n to the distinct points of the set P. Consider our partition of K_n into subgraphs of K_3. Each subgraph of K_3 with at least two edges uniquely determines a circle or a circular arc, which we draw. For each subgraph that consists of a single edge, we draw the line segment that connects the two vertices. The above conditions ensure that the drawings of no two subgraphs have a crossing.

The upper bound for $\sigma_3^1(K_{p,q})$ can be seen as follows. Let $p' = \lceil p/2 \rceil \geq p/2$ and $q' = \lceil q/2 \rceil \geq q/2$. Draw a bipartite graph $K_{2p',2q'} \supset K_{p,q}$ in 3D as follows; see Fig. 3.1b. Let $V(K_{2p',2q'}) = P \cup Q$ be the natural bipartition of its vertices. Fix any family of p' distinct spheres with a common intersection circle. Place the $2q'$ vertices of Q on q' distinct pairs of antipodal points on the circle. Consider a line going through the center of the circle and orthogonal to its plane. Place the $2p'$ vertices of P into p' pairs of distinct intersection points

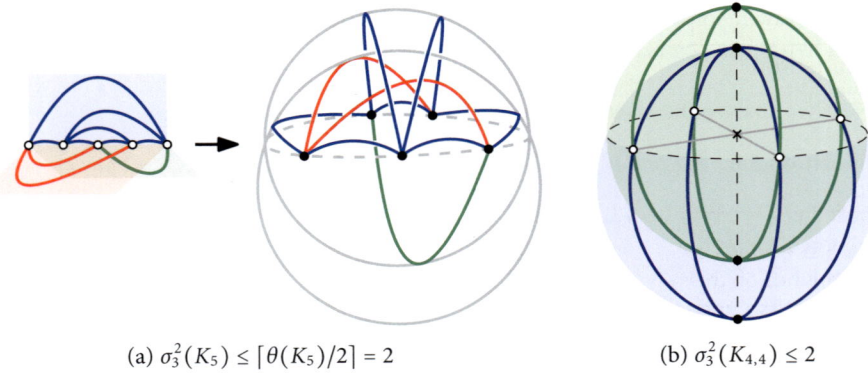

(a) $\sigma_3^2(K_5) \leq \lceil \theta(K_5)/2 \rceil = 2$ (b) $\sigma_3^2(K_{4,4}) \leq 2$

Figure 3.1: Upper bounds for the spherical cover number of complete (bipartite) graphs.

of the line with the circles of the family, the points from each pair belonging to the same sphere. Now each pair of antipodal points in Q together with each pair of cospheric points in P determine a unique circle that contains all these points and provides a drawing of the four edges between them. The union of all these circles is the desired drawing of $K_{2p',2q'}$ onto $p'q'$ circles.   □

We remark that Proposition 3.3 and all the bounds for 3D in this section also hold for higher dimensions.

G	K_n	$K_{p,q}$	references
$\rho_3^1(G)$	$\binom{n}{2}$	$pq - \lfloor \frac{p}{2} \rfloor - \lfloor \frac{q}{2} \rfloor$	[CFL$^+$16, Expl. 10 & 25(c)]
$\rho_3^2(G)$	$\frac{n^2-n}{12} \sim \frac{n^2+5n+6}{6}$	$\left\lceil \frac{\min\{p,q\}}{2} \right\rceil$	[CFL$^+$16, Thm. 12, Expl. 11]
$\sigma_3^1(G)$	$\lfloor \frac{n^2}{8} \rfloor \sim \frac{n^2+5n+6}{6}$	$\left\lceil \frac{pq}{4} \right\rceil \sim \lceil \frac{p}{2} \rceil \lceil \frac{q}{2} \rceil$	Theorem 3.2
$\sigma_3^2(G)$	$\lfloor \frac{n+7}{6} \rfloor \sim \lceil \frac{n}{4} \rceil$	$\left\lceil \frac{pq}{2(p+q-2)} \right\rceil \sim \left\lceil \frac{\min\{p,q\}}{2} \right\rceil$	Theorem 3.1

Table 3.1: Lower and upper bounds on the three-dimensional line, plane, circle, and sphere cover numbers of K_n for any $n \geq 1$ and of $K_{p,q}$ for any $p, q \geq 3$. The cells with only one entry contain tight bounds.

Table 3.1 summarizes the known bounds for the affine cover numbers [CFL$^+$16] and the new bounds for the spherical cover numbers of complete (bipartite) graphs in 3D.

$G = (V, E)$	$|V|$	$|E|$	$|F|$	seg	upp. bd.	arc	lower bd.	upp. bd.
tetrahedron	4	6	4	6	Fig. 3.2a	3	Prop. 3.4(a)	Prop. 3.8(a)
octahedron	6	12	8	9	Fig. 3.3a	3	Prop. 3.4(a)	Prop. 3.8(b)
cube	8	12	6	7	Fig. 3.4a	4	[DESW07, Lem. 5]	Prop. 3.8(c)
dodecahedron	20	30	12	13	Fig. 3.5a	10	[DESW07, Lem. 5]	Prop. 3.8(d)
icosahedron	12	30	20	15	Fig. 3.6a	7	Prop. 3.4(a)	Prop. 3.8(e)

Table 3.2: Bounds on the segment and arc numbers of the platonic graphs. We obtained the lower bounds on the segment number with the help of an integer linear program; see Table 3.4 in Section 3.5. The upper bounds for the segment numbers of the dodecahedron and icosahedron have been established by Schulz [Sch15] and Scherm [Sch16, Fig. 2.1(c)].

graph	ρ_2^1	ρ_3^1	lower bd.	upp. bd.	σ_2^1	σ_3^1	upp. bd.
tetrahedron	6		6	[CFL$^+$16, Expl. 10]	Fig. 3.2a	3	3	Fig. 3.2b
octahedron	9		9	Prop. 3.7(b)	Fig. 3.3a	3	3	Fig. 3.3c
cube	7		7	Prop. 3.7(c)	Fig. 3.4a	4	4	Fig. 3.4d
dodecahedron	9…10	9…10	Prop. 3.7(d)	Fig. 3.5a	5	5	Fig. 3.5d
icosahedron	13…15	13…15	Prop. 3.7(e)	Fig. 3.6a	7	7	Fig. 3.6c

Table 3.3: Bounds on the affine cover numbers ρ_d^l and the spherical cover numbers σ_d^l for platonic graphs. The lower bounds on σ_2^1 and σ_3^1 stem from Proposition 3.4(a).

3.3 Platonic Graphs

In this section we analyze the segment numbers, arc numbers, affine cover numbers, and spherical cover numbers of platonic graphs. We provide upper bounds via the corresponding drawings; see Figures 3.2–3.6.

To bound the spherical cover numbers σ_2^1 and σ_3^1 of the platonic graphs from below, we use a single combinatorial argument—Proposition 3.4(a); see Section 3.4. For the affine cover number ρ_2^1, a similar combinatorial argument fails [CFL$^+$16, Lemma 7(a)]. Therefore, we bound ρ_3^1 (and, hence, also ρ_2^1) from below for each platonic graph individually; see Proposition 3.7. For an overview of our results, see Tables 3.2 and 3.3. We abbreviate every platonic graph by its capitalized initial; for example, C for the cube.

Proposition 3.7. (a) $\rho_3^1(T) \geq 6$; (b) $\rho_3^1(O) \geq 9$; (c) $\rho_3^1(C) \geq 7$; (d) $\rho_3^1(D) \geq 9$; (e) $\rho_3^1(I) \geq 13$.

Proof: (a) Follows from [CFL$^+$16, Ex. 10].

(b) Consider a straight-line drawing of the octahedron O covered by a family $\mathcal{L}$ of ρ lines. Observe that every vertex of the octahedron is adjacent to every other except the opposite vertex. Therefore, no line in $\mathcal{L}$ can cover more than three vertices, otherwise the edges on the line would overlap. Hence, every line covers at most two edges, and these must be adjacent. Moreover, the two end vertices of these length-2 paths cannot be adjacent. Since there are only three pairs of such vertices, at most three lines cover two edges each. Since the octahedron has twelve edges, $\rho \geq 9$.

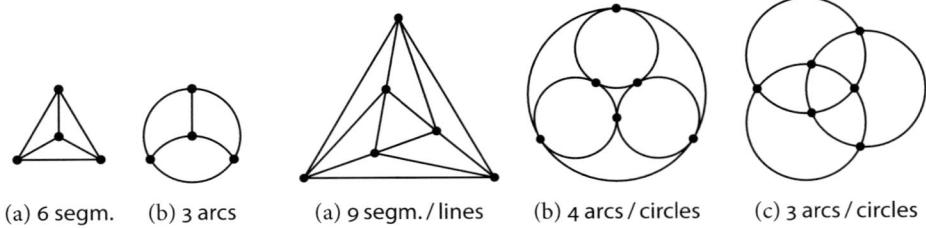

(a) 6 segm.   (b) 3 arcs   (a) 9 segm. / lines   (b) 4 arcs / circles   (c) 3 arcs / circles

Figure 3.2: Drawings of the tetrahedron.
           Figure 3.3: Drawings of the octahedron.

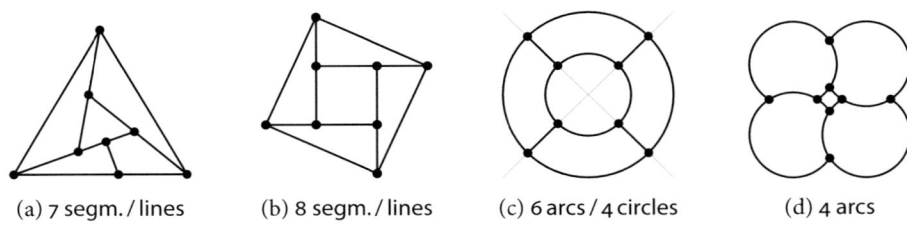

(a) 7 segm. / lines   (b) 8 segm. / lines   (c) 6 arcs / 4 circles   (d) 4 arcs

Figure 3.4: Drawings of the cube.

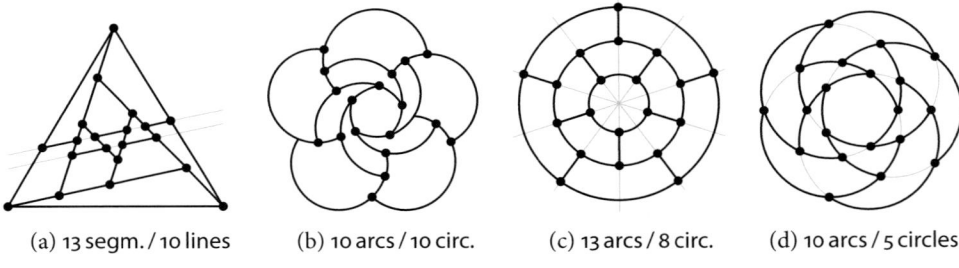

(a) 13 segm. / 10 lines   (b) 10 arcs / 10 circ.   (c) 13 arcs / 8 circ.   (d) 10 arcs / 5 circles

Figure 3.5: Drawings of the dodecahedron: (a), (c)[Sch16]; (b)[Sch15].

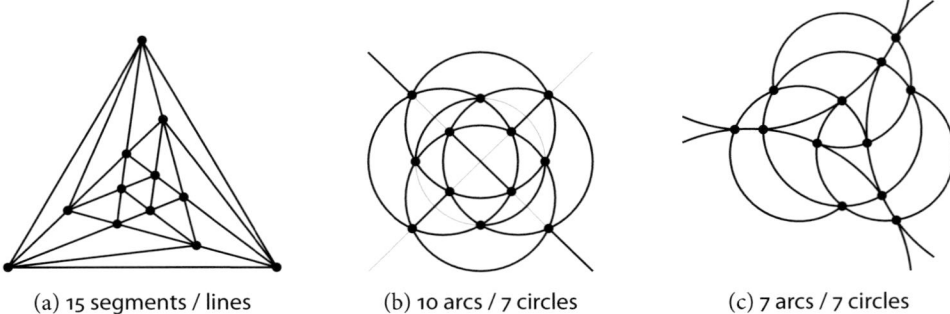

(a) 15 segments / lines   (b) 10 arcs / 7 circles   (c) 7 arcs / 7 circles

Figure 3.6: Drawings of the icosahedron.

(c) Now consider a straight-line drawing of the cube C covered by a family $\mathcal{L}$ of ρ lines. We distinguish two cases.

Assume first that the drawing of the cube lies in a single plane. Each embedding of the cube contains two nested cycles, namely, the boundary of the outer face and the innermost face. We consider three cases depending on the shape of the outer face. (i) If the outer cycle is drawn as the boundary of a (strictly) convex quadrilateral, then none of the lines covering its sides can be used to cover the edges of the innermost cycle, therefore, it needs three additional lines. (ii) If the outer cycle is drawn as the boundary of a non-convex quadrilateral, then we need three additional lines to cover the three edges going from its three convex angles to the innermost cycle. (iii) Now assume that the outer cycle is drawn as a triangle. Then none of the lines covering its sides can be used to cover the edges of the innermost cycle. If this cycle is drawn as a quadrilateral, then we need four additional lines to cover its sides. If the innermost cycle is drawn as a triangle, then we need three lines for the triangle and an additional line to cover the edge incident to the vertex of the innermost cycle which is not a vertex of the triangle. In each of the three cases (i)–(iii), we need at least seven lines to cover the cube.

Now assume that the drawing of the cube is not contained in a single plane. Then its convex hull has (at least) four extreme points. In order to cover the cube, we need at least one pair of intersecting lines of $\mathcal{L}$ for each vertex of the cube and at least three such pairs for each extreme point, that is, at least $4 + 4 \cdot 3 = 16$ pairs of intersecting straight lines in total. So, $\binom{\rho}{2} \geq 16$ and $\rho \geq 7$.

(d) Consider a straight-line drawing of the dodecahedron D covered by a family $\mathcal{L}$ of ρ lines. Again, we distinguish two cases.

Assume first that the drawing of the dodecahedron lies in a single plane. Again we make a case distinction depending on the shape of the outer cycle. (i) If the outer cycle is drawn as the boundary of a convex polygon, let $\mathcal{L}_0 \subseteq L$ be the family of lines that support the edges on the outer cycle. This family consists of at least three lines. None of them covers any of the at most 15 vertices remaining in the interior of the convex polygon. Thus each of these vertices is an intersection point of two lines of $\mathcal{L} \setminus \mathcal{L}_0$. Since $\mathcal{L} \setminus \mathcal{L}_0 \leq \rho - 3$, this family of lines can generate at most $\binom{\rho-3}{2}$ intersection points. Therefore, $\binom{\rho-3}{2} \geq 15$ and, hence, $\rho \geq 9$. (ii) Assume that the outer cycle is drawn as a non-convex quadrilateral. Then the drawing is contained in a convex angle opposite to the reflex angle. To cover the angle sides, we need a family $\mathcal{L}_0$ consisting of at least two lines. None of them covers any of the at least $15 + 1$ vertices remaining in the interior of the angle. Similarly to the previous paragraph, we obtain $\binom{\rho-2}{2} \geq 16$ and, hence, $\rho \geq 9$. (iii) Assume that the outer cycle is drawn as a pentagon P. Since the angle sum of a pentagon is 3π, P has at most two reflex angles, and therefore, at least three convex angles. Each vertex of D drawn as a vertex of a convex angle is an intersection point of (at least) three covering lines, because it has degree 3. There exists an edge e of P such that P is contained in one of the half-planes created by the line ℓ spanned by e (see, for instance, [Mos60]). It is easy to check that ℓ can cover only edge e of the outer face of D. Then the family $\mathcal{L} \setminus \{\ell\}$ covers all edges of G but e. The angles of P incident to e are convex. Let v be a vertex of D drawn as a vertex of a convex angle not incident to e. In order to cover D, we need at least one pair of intersecting lines from $\mathcal{L} \setminus \{\ell\}$ for each vertex of D different from v and at least three such pairs for v, that is, at least $19 + 3 = 22$ pairs of intersecting lines in total. Therefore, $\binom{\rho-1}{2} \geq 22$ and, hence, $\rho \geq 9$. Note that, in each of the three cases (i)–(iii), we have $\rho \geq 9$.

Now assume that the drawing of D is not contained in a single plane. Then its convex hull has (at least) four extreme points. In order to cover D, we need at least one pair of intersecting lines of $\mathcal{L}$ for each vertex of D and at least three such pairs for each extreme point, that is, at least $16 + 4 \cdot 3 = 28$ pairs of intersecting lines in total. Therefore, $\binom{\rho}{2} \geq 28$. But if we have equality then any two lines of $\mathcal{L}$ intersect. So all of them share a common plane or a common point. In the first case the drawing is contained in a single plane; in the second case the family $\mathcal{L}$ cannot cover the drawing. Thus $\binom{\rho}{2} > 28$, and, hence, $\rho \geq 9$.

(e) If the drawing of the icosahedron I is not contained in a single plane, then we can pick four extreme points of the convex hull of the drawing. Each of these points represents a vertex of degree 5, so we need five lines to cover edges incident to this vertex, that is, 20 lines in total, but we have double-counted the lines that go through pairs of the extreme points that we picked. Of these, there are at most $\binom{4}{2} = 6$. Thus we need at least $20 - 6 = 14$ lines to cover the drawing.

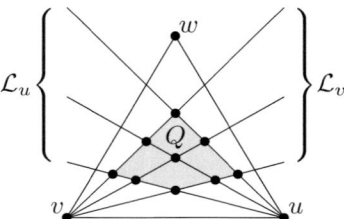

Figure 3.7: The families of lines $\mathcal{L}_u$ and $\mathcal{L}_v$.

Now assume that there exists a straight-line drawing of the icosahedron in a single plane covered by a family $\mathcal{L}$ of twelve lines. Let u, v, w be the vertices of the outer face of I. Clearly, three distinct lines in $\mathcal{L}$ form the triangle uvw. For $s \in \{u, v, w\}$, we denote by $\mathcal{L}_s$ the lines in $\mathcal{L}$ that go through s and do not cover edges of the outer face. Since I is 5-regular, $|\mathcal{L}_s| = \deg(s) - 2 = 3$. Consider the set P of intersection points between the line families $\mathcal{L}_u$ and $\mathcal{L}_v$. The set P lies in the triangle uvw and is bounded by the quadrilateral Q formed by the outer pairs of lines in $\mathcal{L}_v$ and $\mathcal{L}_u$; see Fig. 3.7.

The quadrilateral Q is convex and eight of the nine points in P lie on the boundary of Q, hence, for any line ℓ in $\mathcal{L}_w$, we have $|\ell \cap P| \leq 3$. Observe that $|\ell \cap P| = 3$ implies that ℓ goes through the only point of P that lies in the interior of Q. Thus the lines in $\mathcal{L}_w$ can create at most seven triple intersection points with the lines in $\mathcal{L}_u$ and $\mathcal{L}_v$.

The icosahedron is 5-regular, so all vertices must be placed at the intersection of at least three lines. We need at least nine triple intersection points in order to place all $12 - 3$ inner vertices of the icosahedron—a contradiction. $\qquad \square$

Proposition 3.8. *(a)* arc$(T) \leq 3$; *(b)* arc$(O) \leq 3$; *(c)* arc$(C) \leq 4$; *(d)* arc$(D) \leq 10$; *(e)* arc$(I) \leq 7$.

Proof: For the upper bounds for (a)–(d) see the drawings of the graphs in Figures 3.2b, 3.3c, 3.4d, and 3.5d respectively. While it is easy to see that these drawings are valid, we argue more carefully that the icosahedron does indeed admit a drawing with seven arcs. To

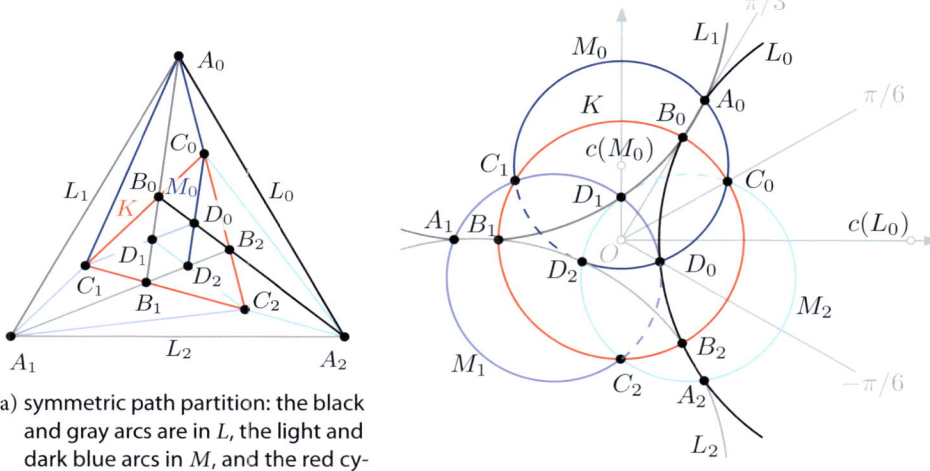

(a) symmetric path partition: the black and gray arcs are in L, the light and dark blue arcs in M, and the red cycle is K

(b) illustration of the proof of Proposition 3.8(e)

Figure 3.8: Bounding the arc number of the icosahedron.

construct the drawing in Fig. 3.6c (for details see Fig. 3.8), we first cover the edges of the icosahedron by seven objects, grouped into a single cycle K and two sets $L = \{L_0, L_1, L_2\}$ and $M = \{M_0, M_1, M_2\}$, where K is a cycle of length 6 and all elements of L and M are simple paths of length 4; see Fig. 3.8a. We identify the paths and cycles with their drawings as arcs and circles. For a set $S \in \{\{K\}, L, M\}$ and a number $i \in \{0, 1, 2\}$, let (d_S, α_{S_i}) be the polar coordinates of the center $c(S_i)$ of the circle of radius r_S that covers arc $S_i \in S$ (see Fig. 3.8b). We set the coordinates and radii as follows:

$$\alpha_K = 0 \qquad\qquad d_K = 0 \qquad\qquad r_K = 1$$

$$\alpha_{L_i} = i \cdot 2\pi/3 \qquad\qquad d_L = (3 + \sqrt{3})/2 \qquad\qquad r_L = \sqrt{5/2 + \sqrt{3}}$$

$$\alpha_{M_i} = \pi/2 + i \cdot 2\pi/3 \qquad\qquad d_M = (3 - \sqrt{3})/2 \qquad\qquad r_M = \sqrt{5/2 - \sqrt{3}}$$

Using the law of cosines, it is easy to compute the intersection points:

$$\begin{aligned}
\{A_i\} &:= L_i \cap L_{i+1} \cap M_i & &\Rightarrow A_i = \left(i \cdot 2\pi/3, (1 + \sqrt{3})/2\right); \\
\{B_i\} &:= L_i \cap L_{i+1} \cap K & &\Rightarrow B_i = (i \cdot 2\pi/3, 1); \\
\{C_i\} &:= M_i \cap M_{i+2} \cap K & &\Rightarrow C_i = (\pi/3 + i \cdot 2\pi/3, 1); \\
\{D_i\} &:= L_i \cap M_i \cap M_{i+1} & &\Rightarrow D_i = \left(\pi/2 + i \cdot 2\pi/3, (\sqrt{3} - 1)/2\right).
\end{aligned}$$

For $i = 0, 1, 2$, let L_i be the larger arc of the covering circle between the points A_i and B_i, let M_i be the larger arc of the covering circle between the points C_{i+1} and D_{i+2} (with indices modulo 3), and let K be the whole unit circle. $\qquad\square$

3.4 Lower Bounds for σ_d^1

Given a graph G, we obtain lower bounds for $\sigma_d^1(G)$ via standard combinatorial characteristics of G in the same way as for the bounds for $\rho_d^1(G)$ [CFL+16]. In particular, we show that $\sigma_d^1(G)$ is bounded from below by the W-separation number $\text{sep}_W(G)$; see Section 2.1 for the definition. Moreover, we prove a general lower bound for $\sigma_d^1(G)$ in terms of the treewidth $\text{tw}(G)$ (see Section 2.1 for the definition) of G, which follows from the fact that graphs with low parameter $\sigma_d^1(G)$ have small separators. This fact is interesting by itself and has yet another consequence: graphs with bounded vertex degree can have a linearly large value of $\sigma_d^1(G)$ (hence, the factor of n in the trivial bound $\sigma_d^1(G) \le m \le n \cdot \Delta(G)/2$ is best possible).

In addition, for any graph G we provide a lower bound with respect to the *linear arboricity* $\text{la}(G)$, that is, the minimum number of linear forests that partition the edge set of G [Har70].

The proofs for the lower bounds are similar to those regarding the affine cover number [CFL+16]. We restate Proposition 3.5 as item (a) to make the following theorem more self-contained.

Theorem 3.3. *For any integer $d \ge 1$ and any graph G with n vertices and m edges, the following bounds hold:*

(a) $\sigma_d^1(G) \ge \text{bw}(G)/2$;

(b) $\sigma_d^1(G) > n/10$ *for almost all cubic graphs with n vertices;*

(c) $\lceil \frac{3}{2}\sigma_d^1(G) \rceil \ge \text{la}(G)$;

(d) $\sigma_d^1(G) \ge \text{sep}_W(G)/2$ *for every $W \subseteq V(G)$;*

(e) $\sigma_d^1(G) \ge \text{tw}(G)/6 - 1$.

Proof: For the proof of (a) see Proposition 3.5.

(b) The claim follows from (a) and from the fact that a random cubic graph on n vertices has bisection width at least $n/4.95$ with probability $1 - o(1)$ [KM93].

(c) Given the drawing of the graph G on $r = \sigma_d^1(G)$ circles, we remove an edge from each of the circles (provided such an edge exists), obtaining at (most) r linear forests. The removed edges we group into (possible, degenerated) pairs, obtaining at most $\lceil r/2 \rceil$ additional linear forests. So, $\text{la}(G) \le r + \lceil r/2 \rceil$.

(d) The proof is similar to Theorem 9(c) in [CFL+16]. The difference of a factor of $1/2$ is due to the fact that a straight line pierces the plane at most once whereas and a circle pierces the hyperplane at most twice.

(e) follows from (d) and the fact that $\text{tw}(G) \le 3k + 2$ if $\text{sep}_W(G) \le k$ for every W with $|W| = 2k + 3$ (see Chapter 2). □

Corollary 3.1. $\sigma_d^1(G)$ *cannot be bounded from above by a function of* $\text{la}(G)$ *or* $v_{\ge 3}(G)$ *or* $\text{tw}(G)$, *where* $v_{\ge 3}(G)$ *is the number of vertices with degree at least 3.*

Proof: la(G): Akiyama et al. [AEH80] showed that, for any cubic graph G, la(G) = 2. On the other hand, $v_{\geq 3}(G) = n$, so $\sigma_3^1(G) > \sqrt{n}$ by Proposition 3.4(a). Theorem 3.3(b) yields an even larger gap.

$v_{\geq 3}(G)$: Let G be the disjoint union of k cycles. Then $v_{\geq 3}(G) = 0$. Clearly, an arrangement A of ℓ circles has at most ℓ^2 vertices. Each cycle of G "consumes" at least two vertices of A or a whole circle, so $\sigma_d^1(G) = \Omega(\sqrt{k})$.

tw(G): Let G be a caterpillar with linearly many vertices of degree 3. Then, tw(G) = 1. On the other hand, by Proposition 3.4(a), we have $\sigma_d^1(G) = \Omega(\sqrt{n})$. $\qquad\square$

Lemma 3.1. *A circular-arc drawing $\Gamma \subset \mathbb{R}$ of a graph G that contains k nested cycles cannot be covered by fewer than k circles.*

Proof: Fix any point inside the closed Jordan curve in Γ that corresponds to the innermost cycle of G. Let ℓ be an arbitrary line through this point. Then ℓ crosses at least twice each of the Jordan curves that correspond to the nested cycles in G. Hence, there are at least $2k$ points where ℓ crosses Γ.

On the other hand, consider any set of r circles whose union covers Γ. Then it is clear that ℓ crosses each of these r circles in at most two points, so there are at most $2r$ points where ℓ crosses Γ. Putting together the two inequalities, we get $r \geq k$ as desired. $\qquad\square$

At last we remark that there are graphs whose σ_3^1-value is a lot smaller than their σ_2^1-value.

Theorem 3.4. *For infinitely many n there is a planar graph G on n vertices with $\sigma_2^1(G) = \Omega(n)$ and $\sigma_3^1(G) = O(n^{2/3})$.*

Proof: We use the same family $(G_i)_{i\geq 1}$ of graphs as Chaplick et al. [CFL$^+$16, Theorem 24(b)] with $G_i = C_3 \times P_i$ and P_i a path with i vertices. Then G_i has $n_i = 3i$ vertices, $\rho_2^1(G_i) = \Omega(n_i)$, and $\rho_3^1(G_i) = O(n_i^{2/3})$. The lower bound on $\sigma_2^1(G_i)$ follows from Lemma 3.1. The upper bound on $\sigma_3^1(G_i)$ follows from Proposition 3.1 for $l = 1$ and $d = 3$, which states that, for any graph G, $\sigma_3^1(G) \leq \rho_3^1(G)$. $\qquad\square$

3.5 An MIP for Estimating the Segment Number

In this section, we exploit a mixed-integer programming formulation for *locally consistent angle assignments* [DETT99], which we define below, to obtain lower bounds on the segment numbers of planar graphs. Our MIP determines a locally consistent angle assignment with the maximum number of π-angles between incident edges. Note that such angle assignments are not necessarily realizable with straight-line edges in the plane. This is why the MIP yields only an upper bound for the number of π-angles—and a lower bound for the segment number. For the platonic graphs, however, it turns out that the bounds are tight; see Tables 3.2 and 3.4.

Let G be a 3-connected graph with fixed embedding given by a set $\mathcal{F}$ of faces and an outer face f_0. Denote the set of vertices $V(G)$ of G as V and the set of edges $E(G)$ of G as E. For any vertex $v \in V$ and any face $f \in \mathcal{F}$, we introduce a fractional variable $x_{v,f} \in (0, 2)$ whose value is intended to express the size of the angle at v in f, divided by π. Thus, $(\pi \cdot x_{v,f})_{v \in V, f \in \mathcal{F}}$ is an angle assignment for G. The following constraints guarantee that the assignment is locally consistent. (For a vertex v and a face f, we write $v \sim f$ to express that v is incident to f.)

$$\sum_{f \sim v} x_{v,f} = 2 \qquad\qquad \text{for each } v \in V;$$

$$\sum_{v \sim f} x_{v,f} = \deg(f) - 2 \qquad\qquad \text{for each } f \in \mathcal{F} \smallsetminus \{f_0\};$$

$$\sum_{v \sim f_0} x_{v,f_0} = \deg(f_0) + 2.$$

For any vertex v, let $L_v = \langle v_1, \ldots, v_k \rangle$ be the list of vertices adjacent to v, in clockwise order as they appear in the embedding. Due to the 3-connectivity of G, any two vertices v_t and v_{t+1} that are consecutive in L_v (and adjacent to v) uniquely define a face $f(v, t)$ incident to v, v_t, and v_{t+1}. For two vertices v_i and v_j with $i < j$, we express the angle $\angle (v_i v v_j)$ as the sum of the angles at v in the faces between v_i and v_j. As shorthand, we use $y_{v,i,j} = \angle (v_i v v_j)/\pi \in (0, 2)$:

$$y_{v,i,j} = \sum_{t=i}^{j-1} x_{v,f(v,t)} \qquad\qquad \text{for each } v \in V, 1 \le i < j \le \deg(v).$$

We want to maximize the number of π-angles between any two edges incident to the same vertex. To this end, we introduce a 0–1 variable $s_{v,i,j}$ for any vertex v and $1 \le i < j \le \deg(v)$. The intended meaning of $s_{v,i,j} = 1$ is that $\angle (v_i v v_j) = \pi$. We add the following constraints to the MIP:

$$\left.\begin{aligned} s_{v,i,j} &\in \{0,1\} \\ s_{v,i,j} &\le y_{v,i,j} \\ s_{v,i,j} &\le 2 - y_{v,i,j} \end{aligned}\right\} \qquad\qquad \text{for each } v \in V, 1 \le i < j \le \deg(v).$$

If $y_{v,i,j} < 1$, the second constraint will force $s_{v,i,j}$ to be 0 and the third constraint will not be effective. If $y_{v,i,j} > 1$, the third constraint will force $s_{v,i,j}$ to be 0, and the second constraint will not be effective. Only if $y_{v,i,j} = 1$ (and $\angle (v_i v v_j) = \pi$), both constraints will allow $s_{v,i,j}$ to be 1. This works because we want to maximize the total number of π-angles between incident edges in a locally consistent angle assignment. To this end, we use the following objective:

$$\text{Maximize} \quad \sum_{v \in V} \sum_{1 \le i < j \le \deg(v)} s_{v,i,j}.$$

Every π-angle between incident edges saves a segment; hence, in any straight-line drawing of G, the number of segments equals the number of edges minus the number of π-angles. In particular, this holds for a drawing that minimizes the number of segments (and simultaneously maximizes the number of π-angles). Thus,

$$\text{seg}(G) = |E| - \text{ang}_\pi(G).$$

Since

$$\text{ang}_\pi(G) \le \sum_{v \in V} \sum_{1 \le i < j \le \deg(v)} s_{v,i,j},$$

graph G	octahedron	cube	dodecahedron	icosahedron
$\text{ang}_\pi(G) \leq$	3	5	17	15
$\text{seg}(G) \quad \geq$	9	7	13	15
variables	60	48	120	180
constraints	137	114	277	395
runtime [s]	0.011	0.009	0.015	0.066

Table 3.4: Upper bounds on the number of π-angles and corresponding lower bounds on the segment numbers of the platonic graphs (except for the tetrahedron) obtained by the MIP and sizes of the MIP formulation for these instances. Running times were measured on a 64-bit machine with 7.7 GB main memory and four Intel i5 cores with 1.90 GHz, using the MIP solver IBM ILOG CPLEX Optimization Studio 12.6.2.

the above relationship provides the lower bound

$$\text{seg}(G) \geq |E| - \sum_{v \in V} \sum_{1 \leq i < j \leq \deg(v)} s_{v,i,j}$$

for the segment number, which can be computed by solving the MIP.

Our MIP has $O(n^3)$ variables and constraints. The experimental results for the platonic graphs (which are 3-connected and thus have a unique planar embedding) are displayed in Table 3.4.

In addition, to check the capabilities of the MIP, we have tested it on a family of triangulations $(G_k)_{k \geq 2}$ constructed by Dujmović et al. [DESW07, Lemma 17]; a variant of the nested-triangles graph (see Fig. 3.9a). The graph G_2 is the octahedron. For $k > 2$, the triangulation G_k is created by recursively nesting a triangle into the innermost triangle of G_{k-1} and connecting its vertices to the vertices of the triangle it was nested into. Note that G_k has $n_k = 3k$ vertices. Dujmović et al. showed a lower bound of $2n_k - 6$ on the segment number of G_k and a tight lower bound of $2n_k - 3$ (see the proof of [DESW07, Lemma 17] and Fig. 3.9a) on the number of segments given the fixed embedding. Figure 3.9b shows the runtime of the MIP in logarithmic scale for the triangulations $G_2, G_3, \ldots, G_8$. As expected, the runtime is (at least) exponential. Interestingly, for each of these (embedded) graphs, our MIP finds a solution with $2n_k - 3$ "segments", thus matching the tight lower bound of Dujmović et al. for the fixed-embedding case.

3.6 Discussions and Open Problems

As mentioned in the introduction, we now show that minimum-line drawings are indeed different from ρ_2^1-optimal drawings. Then we state some open problems regarding affine and spherical cover numbers.

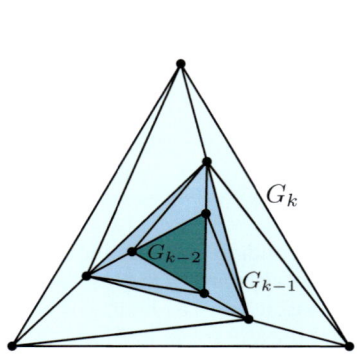

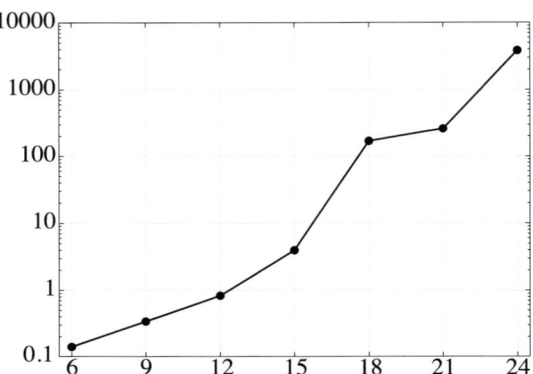

(a) optimal drawing of the triangulation G_k (with $n_k = 3k$ vertices) of Dujmović et al. [DESW07] using $2n_k - 3$ segments

(b) runtime of the MIP applied to the graphs $G_2, G_3, \ldots, G_8$ of Dujmović et al. [DESW07]; the numbers of vertices of the graphs are on the x-axis; the runtime in seconds is on the y-axis; note the log-scale at the y-axis

Figure 3.9: Testing the MIP: instances and runtime.

Example 3.1. Minimum-line drawings are different from ρ_2^1-optimal drawings.

Proof: We provide a graph G with $\rho_2^1(G) = 5$ and $\mathrm{seg}(G) \leq 6$; see Fig. 3.10a. Then we show that every embedding of G on any arrangement of five straight lines consists of at least seven segments.

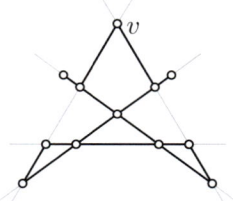

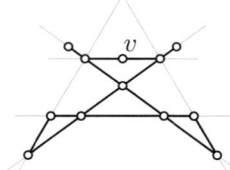

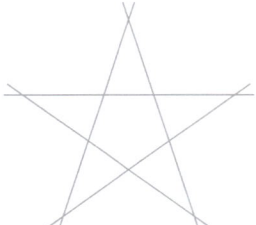

(a) a ρ_2^1-optimal drawing of G on 5 lines and 7 segments

(b) a minimum-line drawing of G on 6 lines and 6 segments

(c) a star-shaped arrangement of 5 straight lines

Figure 3.10: A graph G that shows that ρ_2^1-optimal drawing and minimum-line drawings are indeed different.

Chaplick et al. [CFL+16] defined a vertex of a planar graph to be *essential* if it has degree at least 3 or belongs to a cycle of length 3. They observe that in any drawing of a graph any essential vertex is shared by two edges not lying on the same line. Observe that G has nine essential vertices. Hence, any arrangement of straight lines that cover a drawing of G consists of at least five straight lines (with potentially ten intersection points). Moreover, for the same reason, an arrangement of five straight lines covering a drawing of G must be *simple*, that is, every two straight lines intersect and no three straight lines have a point in common. There is only one such arrangement of five straight lines in the projective plane [Gru72]. This combinatorially unique arrangement is star-shaped; see Fig. 3.10c.

The graph G has three triangles that are attached via one vertex in a chain-like fashion. These triangles can only be embedded into faces of the arrangement; otherwise there would be a triangle that consumes two additional intersection points of the arrangement. Therefore, there is only one way to embed the three triangles on the arrangement, namely on some three consecutive spikes of the star. This forces the degree-2 vertex v (see Fig. 3.10a) to be on a bend (incident to two segments in the drawing) and makes the embedding combinatorially unique. In this embedding of G we have seven segments, but $\mathrm{seg}(G) \leq 6$; see Fig. 3.10b.

Finally, if a graph does not have a drawing with six segments covered by five straight lines in the projective plane, it also does not have one in the Euclidean plane, because we can embed a line arrangement in the Euclidean plane into one in the projective plane preserving the number of segments. So we need at least six straight lines for a drawing with six segments. $\quad\square$

We close with some open problems.

We conjecture that our drawings in Figures 3.5a and 3.6a are optimal. This would mean that $\rho_3^1(D) = 10$ and $\rho_3^1(I) = 15$, but we have no proof for this.

Is there a family $\mathcal{G}$ of graphs such that the affine cover number $\rho_d^l(G)$ of every graph $G \in \mathcal{G}$ can be bounded by a function of the spherical cover number $\sigma_d^l(G)$? For example in the plane (recall that we only consider planar graphs there), $\rho_2^1(G) \in O(n)$, since we can use a different line for each single edge, moreover, according to Proposition 3.4(a) $\sigma_2^1(G) \in \Omega(\sqrt{n})$, therefore, we have that $\rho_2^1(G) \in O(\sigma_2^1(G)^2)$. For the given family of graphs can this relation be tightened? For example, Chaplick et al. [CFL$^+$16, Example 22] showed that there are triangulations for which $O(\sqrt{n})$ lines suffice. It would be even more interesting to find families of graphs where there is an asymptotic difference between the two cover numbers.

We have already seen that $\sigma_3^2(K_n)$ grows asymptotically more slowly than $\rho_3^2(K_n)$. Is there a family of planar graphs where σ_2^1 grows asymptotically more slowly than ρ_2^1?

Chaplick et al. [CFL$^+$16] showed that the hierarchy of affine cover numbers collapses in the following sense: For every graph G, for every integer $d > 3$, and for every integer l with $1 \leq l \leq d$, it holds that $\rho_d^l(G) = \rho_3^l(G)$. The proof of this fact is based on affine maps, which transform planes into planes, but not spheres into spheres, so we don't know whether the hierarchy of spherical cover numbers collapses, too.

On Arrangements of Orthogonal Circles

For the purpose of this chapter, an *arrangement* is a (finite) collection of curves such as lines or circles in the plane. The study of arrangements has a long history; for example, Grün-baum [Gru72] studied arrangements of lines in the projective plane. Arrangements of circles and other closed curves have also been studied extensively [AAS03, ALPS01, FS18, KM14, Pin02]. An arrangement is *simple* if no point of the plane belongs to more than two curves and every two curves intersect. A *face* of an arrangement $\mathcal{A}$ in the projective or Euclidean plane P is a connected component of the subdivision induced by the curves in $\mathcal{A}$, that is, a face is a component of $P \smallsetminus \bigcup \mathcal{A}$.

For a given type of curves, people have investigated the maximum number of faces that an arrangement of such curves can form. In 1826, Steiner [Ste26] showed that a simple arrangement of straight lines can have at most $\binom{n}{2} + \binom{n}{1} + \binom{n}{0}$ faces while an arrangement of circles can have at most $2\left(\binom{n}{2} + \binom{n}{0}\right)$ faces.

Alon et al. [ALPS01] and Pinchasi [Pin02] studied the number of *digonal* faces, that is, faces that are bounded by two edges, for various kinds of arrangements of circles. For example, any arrangement of n unit circles has $O(n^{4/3} \log n)$ digonal faces [ALPS01] and at most $n + 3$ digonal faces if every pair of circles intersects [Pin02], whereas arrangements of circles with arbitrary radii have at most $20n - 2$ digonal faces if every pair of circles intersects [ALPS01].

The same arrangements can, however, have quadratically many *triangular* faces, that is, faces that are bounded by three edges. A lower bound example with quadratically many triangular faces can be constructed from a simple arrangement $\mathcal{A}$ of lines by projecting it on a sphere (disjoint from the plane containing $\mathcal{A}$) and having each line become a great circle. This is always possible since the line arrangement is simple; for more details see [Fel04, Section 5.1]. In this process we obtain $2p_3$ triangular faces, where p_3 is the number of triangular faces in the line arrangement. The great circles on the sphere can then be transformed into a circle arrangement in a different plane using the stereographic projection. This gives rise to an arrangement of circles with $2p_3$ triangular faces in this plane. Füredi and Palásti [FP84] provided simple line arrangements with $n^2/3 + O(n)$ triangular faces. With the argument above, this immediately yields a lower bound of $2n^2/3 + O(n)$ on the number of triangular faces of arrangements of circles. Felsner and Scheucher [FS18] showed that this lower bound is tight by proving that an arrangement of *pseudocircles* (that is, closed curves that can intersect at most twice and no point belongs to more than two curves) can have at most $2n^2/3 + O(n)$ triangular faces.

One can also specialize circle arrangements by fixing an angle (measured as the angle between the two tangents at either intersection point) at which each pair of intersecting circles intersect; this was recently discussed by Eppstein [Epp18a]. An arrangement of circles in which each intersecting pair intersect at a right angle is called *orthogonal*. Note that in any arrangement of orthogonal circles no two circles can touch and no three circles can intersect at the same point.

The main result of this chapter is that arrangements of n orthogonal circles have at most $14n$ intersection points and at most $15n + 2$ faces; see Theorem 4.1. This is different from arrangements of orthogonal circular arcs, which can have quadratically many quadrangular faces; see the arcs inside the blue square in Figure 4.5. In Section 4.3 we also consider small (that is, digonal and triangular) faces and provide bounds on the number of such faces in arrangements of orthogonal circles.

Given a set of geometric objects, their *intersection graph* is a graph whose vertices correspond to the objects and whose edges correspond to the pairs of intersecting objects. Restricting the geometric objects to a certain shape restricts the class of graphs that admit a representation with respect to this shape. For example, graphs represented by disks in the Euclidean plane are called *disk intersection graphs*. The special case of *unit disk graphs*—intersection graphs of unit disks—has been studied extensively. Recognition of such graphs as well as many combinatorial problems restricted to these graphs such as coloring, independent set, and domination are all NP-hard [CCJ90]; see also the survey of Hliněný and Kratochvíl [HK01]. Instead of restricting the radii of the disks, people have also studied restrictions of the type of intersection. If the disks are only allowed to touch, the corresponding graphs are called *coin graphs*. Koebe's classical result says that the coin graphs are exactly the planar graphs. If all coins have the same size, the represented graphs are called *penny graphs*. These graphs have been studied extensively, too [DP11, CFFP11, Epp18b]. For example, they are NP-hard to recognize [BK98, DETT99].

As with the arrangements above, we again consider a restriction on the intersection angle. We define the *orthogonal circle intersection graphs* as the intersection graphs of arrangements of orthogonal circles. In Section 4.4, we investigate properties of these graphs. For example, similar to the proof of our linear bound on the number of intersection points for arrangements of orthogonal circles (Theorem 4.1), we observe that such graphs have only a linear number of edges.

We also consider *orthogonal unit circle intersection graphs*, that is, orthogonal circle intersection graphs with a representation that consists only of unit circles. We show that these graphs are a proper subclass of penny graphs. It is NP-hard to recognize penny graphs [EW96]. We modify the NP-hardness proof of Di Battista et al. [DETT99, Section 11.2.3], which uses the *logic engine*, to obtain the NP-hardness of recognizing orthogonal unit circle intersection graphs (Theorem 4.4).

4.1   **Preliminary Results**

Circles crossing at right angles are most commonly known from *inversive geometry*. Inversive geometry is the study of properties of geometric objects that are preserved after a certain type of transformation called *inversion*. To define inversion more precisely let us define some helpful notation first. For any circle γ, let $C(\gamma)$ be its center and let $r(\gamma)$ be its radius. The inversion with respect to α is a mapping that maps any point $P \neq C(\alpha)$ to a point P' on the ray $C(\alpha)P$ so that $|C(\alpha)P'| \cdot |C(\alpha)P| = r(\alpha)^2$. Inversion maps each circle not passing through $C(\alpha)$ to another circle and a circle passing through $C(\alpha)$ to a line; see Figure 4.1. Thus, one property that inversion preserves is that every circle in the preimage is still a circle in the image (we consider each line a circle of infinite radius). Another useful property of

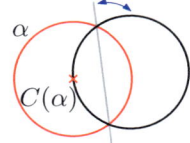

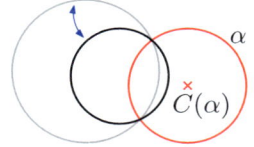

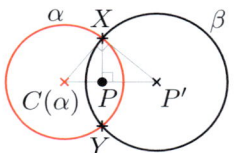

(a) a circle passing through $C(\alpha)$ is mapped to a line (and vice versa)

(b) a circle not passing through $C(\alpha)$ is mapped to another circle

(c) constructing the inversion P' of a point P w.r.t. α via a circle β orthogonal to α

Figure 4.1: Examples of inversion.

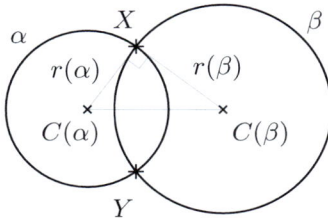

Figure 4.2: Circles α and β are orthogonal if and only if $\triangle XC(\alpha)C(\beta)$ is right-angled.

inversion, is that it preserves angles. So we can transform geometric objects into different possibly more convenient settings where the underlying properties are still preserved. In fact, some difficult problems in geometry become much easier to solve after an inversion is applied [Dör65, Ogi69].

Let us now define *orthogonal circles*. An angle at which two circles intersect is the angle between the two tangents to each of the circles at an intersection point. Two circles intersecting at a right angle are called *orthogonal*. The following observation follows from the Pythagorean theorem.

Observation 4.1. *Let α and β be two circles. Then α and β are orthogonal if and only if $r(\alpha)^2 + r(\beta)^2 = |C(\alpha)C(\beta)|^2$; see Figure 4.2.*

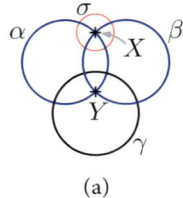
(a)

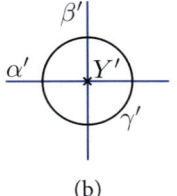
(b)

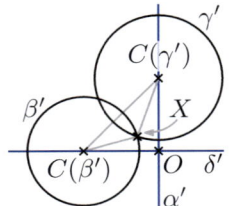

Figure 4.3: (a) Three pairwise intersecting circles, the red inversion circle is centered at X; (b) image of the inversion.

Figure 4.4: Illustration for Lemma 4.2.

In addition we note the following.

Observation 4.2. *Given a pair of orthogonal circles, the tangent to one circle at one of the intersection points goes through the center of the other circle; see Figure 4.2. In particular, a line is orthogonal to a circle if the line goes through the center of the circle.*

Inversion and orthogonal circles are closely related. For example, in order to construct the image P' of some point P that lies inside the inversion circle α, consider the intersection points X and Y of α and the line that is orthogonal to the line through $C(\alpha)$ and P in P; see Figure 4.1c. The point P' then is simply the center of the circle β that is orthogonal to α and goes through X and Y. This follows from the similarity of the orthogonal triangles $\triangle C(\alpha)XP'$ and $\triangle C(\alpha)XP$. Using inversion we can easily show several properties of orthogonal circles.

Lemma 4.1 ([Ogi69]). *In an arrangement of orthogonal circles there cannot be four pairwise orthogonal circles.*

Proof: Assume that there are four pairwise orthogonal circles α, β, γ, and δ. Let X and Y be the intersection points of α and β. Consider the inversion with respect to a circle σ centered at X. The images of α and β are orthogonal lines α' and β' that intersect at Y', which is the image of Y; see Figure 4.3. The image of γ is a circle γ' centered at Y' but so is the image δ' of δ. Thus γ' and δ' are either disjoint or equal, but not orthogonal to each other, a contradiction. □

Lemma 4.2. *In an arrangement of orthogonal circles there cannot be two pairs of circles such that each circle of one pair is orthogonal to each circle of the other pair and the circles within the pairs are not orthogonal.*

Proof: Assume there are two pairs (α, β) and (γ, δ) of circles such that the circles within each pair do not intersect each other and each circle of one pair intersects both circles of the other pair. Consider an inversion via a circle σ centered at one of the intersection points of the circles α and δ. In the image they will become lines α' and δ'. The image β' of the circle β must intersect δ' but not α', therefore, its center must lie on the line δ' and it should be to one side of the line α'; see Figure 4.4. Similarly the center of the image γ' of the circle γ must lie on the line α' and γ' should be to one side of the line δ'. Shift the drawing so that the intersection of α' and δ' is at the origin O and observe that the triangle $\triangle C(\beta')OC(\gamma')$

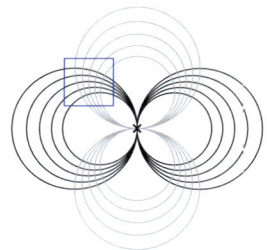
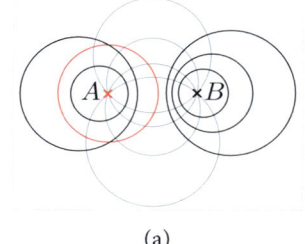
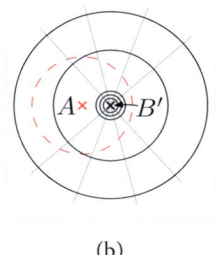

(a)   (b)

Figure 4.5: Apollonian circles consisting of two parabolic pencils of circles.

Figure 4.6: (a) Apollonian circles consisting of an elliptic (in gray) and hyperbolic (in black) pencil of circles; (b) its inversion via a circle centered at A (in red).

is orthogonal, where $C(\beta')$ and $C(\gamma')$ are the centers of the circles β' and γ'. Let X be the intersection point of these circles that is closer to the origin. This point X is contained in the triangle $\triangle C(\beta')OC(\gamma')$. Therefore the triangle $\triangle C(\beta')XC(\gamma')$ cannot be orthogonal—a contradiction. $\qquad\square$

A *pencil* is a family of circles who share a certain characteristic. In a *parabolic* pencil all circles have one point in common, and thus are all tangent to each other; see Figure 4.5. In an *elliptic* pencil all circles go through two given points; see the gray circles in Figure 4.6a. In a *hyperbolic* pencil all circles are orthogonal to a set of circles that go through two given points, that is, to some elliptic pencil; see the black circles in Figure 4.6a.

For an elliptic pencil whose circles share two points A and B and the corresponding hyperbolic pencil, the circles in the hyperbolic pencil possess several properties useful for our purposes [Ogi69]. Their centers are collinear and they consist of non-intersecting circles that form two nested structures of circles, one containing A, the other one containing B in its interior; see Figure 4.6a.

Two pencils of circles such that each circle in one pencil is orthogonal to each circle in the other are called *Apollonian circles*. There can be two such combinations of pencils, that is, one with two parabolic pencils and one with an elliptic and a hyperbolic pencil. We focus on the latter since such Apollonian circles contain arbitrarily large arrangements of orthogonal circles, that is, two orthogonal circles from the elliptic pencil and arbitrary many circles from the hyperbolic pencil. Equivalently, such Apollonian circles are an inversion image of a family of concentric circles centered at some point X and concurrent lines passing through X; see Figure 4.6b. We use this equivalence in the next proof.

Lemma 4.3. *Three circles such that one is orthogonal to the two others belong to the same family of Apollonian circles. Two sets of circles such that each circle in one set is orthogonal to each circle in the other set and each set has at least two circles belong to the same family of Apollonian circles. In particular, if the two sets form an arrangement of orthogonal circles, then the set belonging to the elliptic pencil can contain at most two circles.*

Proof: Consider three circles such that one is orthogonal to two others. If all three are pairwise orthogonal, then their inversion via a circle centered at one of their intersection points (see

Figure 4.3a) is two perpendicular lines and a circle centered at their intersection point (see Figure 4.3b), therefore, they belong to the same family of Apollonian circles. If two circles do not intersect, then by [Ogi69, Theorem 13], it is always possible to invert them into two concentric circles. Since inversion preserves angles, the image of the third circle must be orthogonal to both concentric circles and therefore it must be a straight line passing through the center of both circles. Therefore, the three circles belong to the same family of Apollonian circles.

Consider now two sets S_1 and S_2 of circles such that each circle in one set is orthogonal to each circle in the other set and each set has at least two circles. By Lemma 4.2 there must be two circles α and β in one of the sets, say S_1, that are orthogonal. Consider an inversion via a circle σ centered at one of the intersection points X of the circles α and β. In the image they will become orthogonal lines α' and β' intersecting at a point Y. Because inversion preserves angles, the image of each circle in S_2 is a circle centered at Y. Since S_2 contains at least two circles, the image of each circle in S_1 must be orthogonal to two circles centered at Y, therefore, it must be a straight line passing through Y. Thus, the circles in S_1 and S_2 belong to the same family of Apollonian circles and, if S_1 and S_2 form an arrangement of orthogonal circles, then the set S_1, whose circles belong to the elliptic pencil, can contain at most two circles.   □

4.2   Bounding the Number of Faces

Let $\mathcal{A}$ be an arrangement of orthogonal circles in the plane. By a slight abuse of notation, we will say that a circle α *contains* a geometric object o and mean that the disk bounded by α contains o. We say that a circle $\alpha \in \mathcal{A}$ is *nested* in a circle $\beta \in \mathcal{A}$ if α is contained in β. We say that a circle $\alpha \in \mathcal{A}$ is nested *consecutively* in a circle $\beta \in \mathcal{A}$ if α is nested in β and there is no other circle $\gamma \in \mathcal{A}$ such that α is nested in γ and γ is nested in β. Consider a subset $S \subseteq \mathcal{A}$ of maximum cardinality such that for each pair of circles one is nested in the other. The innermost circle α in S is called a *deepest* circle in $\mathcal{A}$; see Figure 4.7.

Lemma 4.4. *Let α be a circle, and let S be a set of circles orthogonal to α. If S does not contain nested circles and each circle in S has radius at least $r(\alpha)$, then $|S| \leq 6$. Moreover, if $|S| = 6$, then all circles in S have radius $r(\alpha)$ and α is contained in the union of the circles in S.*

Proof: Consider any two circles β and γ in S. Since $r(\beta) \geq r(\alpha)$ and $r(\gamma) \geq r(\alpha)$, the edge $C(\beta)C(\gamma)$ is the longest edge of the triangle $\triangle C(\beta)C(\alpha)C(\gamma)$; see Figure 4.8. So the angle $\angle C(\beta)C(\alpha)C(\gamma)$ is at least $\pi/3$. Thus, $|S| \leq 6$.

Moreover, if $|S| = 6$ then, for each pair of circles β and γ in S that are consecutive in the circular ordering of the circle centers around $C(\alpha)$, it holds that $\angle C(\beta)C(\alpha)C(\gamma) = \pi/3$. This is only possible if $r(\beta) = r(\gamma) = r(\alpha)$. Thus, all the circles in S have radius $r(\alpha)$ and α is contained in the union of the circles in S; see Figure 4.9b.   □

Theorem 4.1. *Every arrangement of n orthogonal circles has at most $14n$ intersection points and $15n + 2$ faces.*

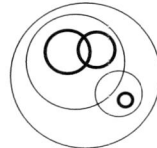

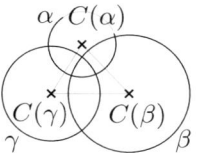

Figure 4.7: Deepest circles in bold.

Figure 4.8: $\angle C(\beta)C(\alpha)C(\gamma) \geq \pi/3$.

The above theorem (whose formal proof is at the end of the section) follows from the fact that any arrangement of orthogonal circles contains a circle α with at most seven *neighbors* (that is, circles that are orthogonal to α).

Lemma 4.5. *Every arrangement of orthogonal circles has a circle that is orthogonal to at most seven other circles.*

Proof: If no circle is nested within any other, Lemma 4.4 implies that the smallest circle has at most six neighbors, and we are done.

So, among the deepest circles in $\mathcal{A}$, consider a circle α with the smallest radius. Note that α is nested in at least one circle. Let β be a circle such that α and β are consecutively nested. Denote the set of all circles in $\mathcal{A}$ that are orthogonal to α but not to β by S_α. All circles in S_α are nested in β. Since α is a deepest circle, S_α contains no nested circles; see Figure 4.9a. Since the radius of every circle in S_α is at least $r(\alpha)$, Lemma 4.4 ensures that S_α contains at most six circles. Given the structure of Apollonian circles (Lemma 4.3), there can be at most two circles that intersect both α and β. This together with Lemma 4.4 immediately implies that α cannot be orthogonal to more than eight circles. In the following we show that there can be at most seven such circles.

If there is only one circle intersecting both α and β, then α is orthogonal to at most seven circles in total, and we are done.

Otherwise, there are two circles orthogonal to both α and β. Let these circles be γ_1 and γ_2. We assume that S_α contains exactly six circles. Hence, by Lemma 4.4, all circles in S_α have radius $r(\alpha)$. Let $S_\alpha = (\delta_0, \ldots, \delta_5)$ be ordered clockwise around α so that every two circles δ_i and δ_j with $i \equiv j + 1 \bmod 6$ are orthogonal.

Let X and Y be the intersection points of γ_1 and γ_2; see Figure 4.9a. Note that, by the structure of Apollonian circles, one of the intersection points, say X, must be contained inside α, whereas the other intersection point Y must lie in the exterior of β. Since the circles in S_α are contained in β, none of them contains Y. Further, no circle δ_i in S_α contains X, as otherwise the circles δ_i, α, γ_1, and γ_2 would be pairwise orthogonal, contradicting Lemma 4.1. Recall that, by Lemma 4.4, α is contained in the union of the circles in S_α. Since X is not contained in this union, γ_1 intersects two different circles δ_i and δ_j, and γ_2 intersects two different circles δ_k and δ_l. Note that γ_1 and γ_2 cannot intersect the same circle ε in S_α, because ε, α, γ_1, and γ_2 would be pairwise orthogonal, contradicting Lemma 4.1. Therefore, the indices i, j, k, and l are pairwise different.

We now consider possible values of the indices i, j, k, and l, and show that in each case we get a contradiction to Lemma 4.1 or Lemma 4.2. If $j \equiv i + 1 \bmod 6$, then γ_1, α, δ_i, and δ_j would be pairwise orthogonal, contradicting Lemma 4.1; see Figure 4.9b. If $j \equiv i + 2 \bmod 6$, then

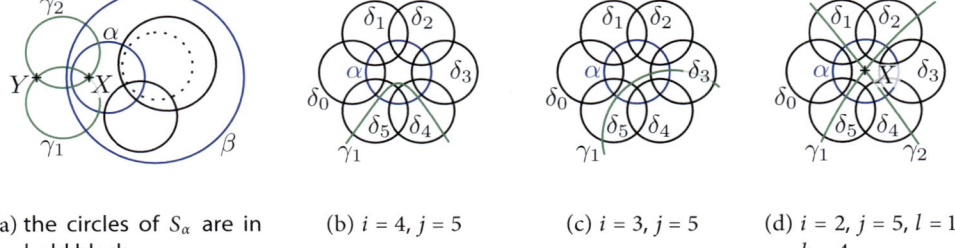

(a) the circles of S_α are in bold black

(b) $i = 4, j = 5$

(c) $i = 3, j = 5$

(d) $i = 2, j = 5, l = 1, k = 4$

Figure 4.9: Illustrations to the proof of Lemma 4.5.

$\gamma_1, \delta_i, \delta_{i+1}$, and δ_j would form an induced C_4 in the intersection graph; see Figure 4.9c. This would contradict Lemma 4.2. If $j \equiv i+3 \mod 6$ and $k \equiv l+3 \mod 6$, then either $k \equiv i+1 \mod 6$ or $i \equiv l + 1 \mod 6$; see Figure 4.9d. W.l.o.g., assume the latter and observe that then $\gamma_2, \delta_i, \gamma_1,$ δ_l would form an induced C_4, again contradicting Lemma 4.2.

We conclude that S_α contains at most five circles. Together with γ_1 and γ_2, at most seven circles are orthogonal to α.   □

Using the lemma above and Euler's formula, we now can prove Theorem 4.1.

Proof of Theorem 4.1: Let $\mathcal{A}$ be an arrangement of orthogonal circles. By Lemma 4.5, $\mathcal{A}$ contains a circle α orthogonal to at most seven circles. The circle α yields at most 14 intersection points. By induction, the whole arrangement has at most $14n$ intersection points.

Consider the planarization G' of $\mathcal{A}$, and let $n', m', f',$ and c' denote the numbers of vertices, edges, faces, and connected components of G', respectively. Since every vertex in the planarization corresponds to an intersection, the resulting graph is 4-regular and therefore $m' = 2n'$. By Euler's formula, we obtain $f' = n'+1+c'$. This yields $f' \le 15n+1$ since $n' \le 14n$ and $c' \le n$.   □

4.3   Bounding the Number of Small Faces

In the following we study the number of faces of each type, that is, the number of digonal, triangular, and quadrangular faces. We begin with some notation. Let $\mathcal{A}$ be an arrangement of orthogonal circles in the plane. Let S be some subset of the circles of $\mathcal{A}$. A face in S is called a region in $\mathcal{A}$ formed by S; see for instance Figure 4.10. Note that each face of $\mathcal{A}$ is also a region.

Let s be the region formed by some circular arcs $a_1, a_2, \ldots, a_k$ enumerated in counterclockwise order around s. For an arc a_i with $i \in \{1, \ldots, k\}$, let α be the circle that supports a_i. If $C(\alpha) = (x_\alpha, y_\alpha)$ is the center of α and $r(\alpha)$ its radius, we can write α as $\{C(\alpha) + r(\alpha)(\cos t, \sin t) : t \in [0, 2\pi]\}$. Let u and v be the endpoints of a_i so that we meet u first when we traverse s counterclockwise when starting outside of a_i. Let $u = C(\alpha) + r(\alpha)(\cos t_1, \sin t_1)$ and $v = C(\alpha) + r(\alpha)(\cos t_2, \sin t_2)$. We say that the region s *subtends* an angle in the circle α of size $\angle(s, a_i) = t_2 - t_1$ with respect to the arc a_i. Note that $\angle(s, a_i)$ is negative if a_i forms a concave side of s. If the circle α forms only one side

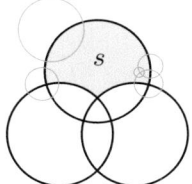

Figure 4.10: Region s is a face in the arrangement of the bold circles.

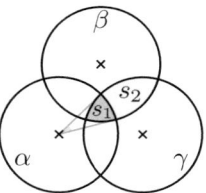

Figure 4.11: Angles subtended by the regions s_1 and s_2 in the circle α; $\angle(s_1, \alpha) = -\angle(s_2, \alpha)$.

of the region s, then we just say that the region s *subtends* an angle in the circle α of size $\angle(s, \alpha) = t_2 - t_1$. Moreover, if s is a digonal region, that is, it is formed by only two circles α and β, then we simply say that β subtends an angle of $\angle(\beta, \alpha) = t_2 - t_1$ in α to mean $\angle(s, \alpha)$.

By *total angle* we denote the sum of subtended angles by s with respect to all the arcs that form its sides, that is, $\sum_{i=1}^{k} \angle(s, a_i)$.

We now give an upper bound on the number of digonal and triangular faces in an arrangement $\mathcal{A}$ of n orthogonal circles. The tool that we utilize in this section is the Gauss–Bonnet formula [Wei19] which, in the restricted case of orthogonal circles in the plane, states that, for every region s formed by some circular arcs $a_1, a_2, \ldots, a_k$, it holds that

$$\sum_{i=1}^{k} \angle(s, a_i) + \frac{k\pi}{2} = 2\pi.$$

This formula implies that each digonal or triangular face subtends a total angle of size π and of size $\pi/2$, respectively. Thus, we obtain the following bounds.

Theorem 4.2. *Every arrangement of n orthogonal circles has at most $2n$ digonal faces and at most $4n$ triangular faces.*

Proof: Because faces do not overlap, each digonal or triangular face uses a unique convex arc of a circle bounding this face. Therefore, the sum of angles subtended by digonal or triangular faces formed by the same circle must be at most 2π. Analogously, the sum of total angles over all digonal or triangular faces cannot exceed $2n\pi$. By the Gauss–Bonnet formula each digonal or triangular face subtends a total angle of size π or $\pi/2$, respectively. This gives an upper bound of $2n$ on the number of digonal faces and an upper bound of $4n$ on the number of triangular faces. □

Theorem 4.2 can be generalized to all convex orthogonal closed curves since the Gauss–Bonnet formula does not require curves to be circular. In contrast to this, for example, a grid made of axis parallel rectangles has quadratically many quadrangular faces. This makes circles a special subclass of convex orthogonal closed curves.

The Gauss–Bonnet formula does not help us to get an upper bound on the number of quadrangular faces. Because each triangular or quadrangular face consists of either three circles such that one is orthogonal to two others or two pairs of circles such that each circle in one pair is orthogonal to each circle in the other pair, we obtain the following observation from Lemma 4.3.

Observation 4.3. *In any arrangement of orthogonal circles, each triangular and each quadrangular face is formed by Apollonian circles.*

Using Observation 4.3, however, it is possible to restrict the types of quadrangular faces to several shapes and obtain bounds on the number of faces of each type. Apart from being interesting in its own right, such a bound also provides a bound on the total number of faces in an arrangement of orthogonal circles. Namely, since the average degree of a face in an arrangement of orthogonal circles is 4, a bound on the number of faces of degree at most 4 gives a bound on the number of all faces in the arrangement (via Euler's formula). Unfortunately, the bound on the number of quadrangular faces that we achieved was $17n$ and thus higher than the bound $15n + 2$ that we now have for the number of *all* faces in an arrangement of n orthogonal circles.

4.4   Intersection Graphs of Orthogonal Circles

Given an arrangement $\mathcal{A}$ of orthogonal circles, consider its *intersection* graph, which is the graph with vertex set $\mathcal{A}$ that has an edge between any pair of intersecting circles in $\mathcal{A}$. Lemmas 4.1 and 4.2 imply that such a graph does not contain any K_4 and any induced C_4. We show that such graphs can be non-planar (Lemma 4.6), then we bound their edge density (Theorem 4.3), and finally we consider the intersection graphs arising from orthogonal *unit* circles (Theorem 4.4).

Lemma 4.6. *For every n, there is an intersection graph of orthogonal circles that contains K_n as a minor. The representation uses circles of three different radii.*

Proof: Let a *chain* be an arrangement of orthogonal circles whose intersection graph is a path. We say that two chains C_1 and C_2 *cross* if two disjoint circles α and β of one chain, say C_1, are orthogonal to the same circle γ of the other chain C_2; see Figure 4.12a (left). If two chains cross, their paths in the intersection graph are connected by two edges; see the dashed edges in Figure 4.12a (right).

Consider an arrangement of n rectilinear paths embedded on a grid where each pair of curves intersect exactly once; see the inset in Figure 4.12b. We convert the arrangement of paths into an arrangement of chains such that each pair of chains crosses; see Figure 4.12b. Now consider the intersection graph of the orthogonal circles in the arrangement of chains. If we contract each path in the intersection graph that corresponds to a chain, we obtain K_n. □

Next, we discuss the density of orthogonal circle intersection graphs. Gyárfás et al. [GHS02] have shown that any C_4-free graph on n vertices with average degree at least a has *clique number* (that is, the number of vertices in the maximum clique of the graph) at least $a^2/(10n)$. Due to Lemma 4.1, we know that orthogonal circle intersection graphs have clique number at most 3. Thus, their average degree is bounded from above by $\sqrt{30n}$, leading to at most $\sqrt{7.5}n^{\frac{3}{2}}$ edges in total. However, Lemma 4.5 implies the following stronger bound.

Theorem 4.3. *The intersection graph of a set of n orthogonal circles has at most $7n$ edges.*

Proof: The geometric representation of an orthogonal circle intersection graph is an arrangement of orthogonal circles. By Lemma 4.5, an arrangement of n orthogonal circles always has

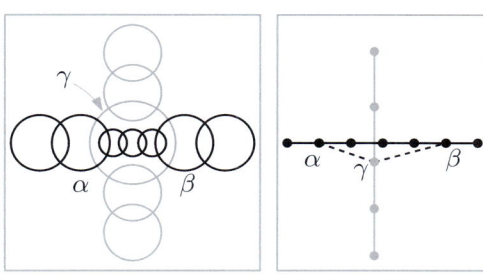
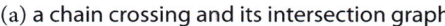
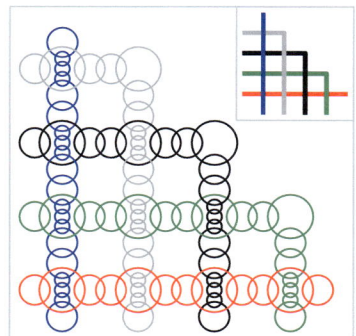

(a) a chain crossing and its intersection graph

(b) pairwise intersecting paths (see inset) and the corresponding chains in an orthogonal circle representation

Figure 4.12: Construction of an orthogonal circle intersection graph that contains K_n as a minor (here $n = 5$).

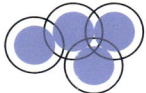

(a) all orthogonal unit circle intersection graphs are penny graphs

(b) penny graphs that aren't orthogonal unit circle intersection graphs

Figure 4.13: Penny graphs vs. orthogonal unit circle intersection graphs.

a circle orthogonal to at most seven circles. Therefore, the corresponding intersection graph always has a vertex of degree at most seven. Thus, it has at most $7n$ edges. □

We also consider a natural subclass of orthogonal circle intersection graphs, the orthogonal *unit* circle intersection graphs. Recall that these are orthogonal circle intersection graphs with a representation that consists of unit circles only. As Figure 4.13a shows, every representation of an orthogonal unit circle intersection graph can be transformed (by scaling each circle by a factor of $\sqrt{2}/2$) into a representation of a *penny graph*, that is, a contact graph of equal-size disks. Hence, every orthogonal unit circle intersection graph is a penny graph – whereas the converse is not true. For example, C_4 or the 5-star are penny graphs but not orthogonal unit circle intersection graphs (see Figure 4.13b).

Orthogonal unit circle intersection graphs being penny graphs implies that they inherit the properties of penny graphs, e.g., their maximum degree is at most six and their edge density is at most $\lfloor 3n - \sqrt{12n - 6} \rfloor$, where n is the number of vertices [PA95, Theorem 13.12, p. 211]. Because triangular grids are orthogonal unit circle intersection graphs, this upper bound is tight.

4.5   Recognizing Orthogonal Unit Circle Intersection Graphs

As it turns out, orthogonal unit circle intersection graphs share another feature with penny graphs: their recognition is NP-hard. The hardness of penny-graph recognition can be shown using the *logic engine* [DETT99, Section 11.2], which simulates an instance of the Not-All-Equal-3-Sat (NAE3SAT) problem. We establish a similar reduction for the recognition of orthogonal unit circle intersection graphs.

In this section, we show how to realize the *logic engine* with orthogonal unit circle intersection graphs. The logic engine simulates the Not-All-Equal-3-Sat (NAE3SAT) problem where a set C of clauses each containing three literals from a set of Boolean variables U is given and the question is to find a truth assignment to the variables so that each clause contains at least one true literal and at least one false literal.

Figure 4.14: Orthogonal unit circle representation of the universal part of the logic engine; only half of the drawing is present, the other half is symmetric.

Theorem 4.4.   *It is NP-hard to recognize orthogonal unit circle intersection graphs.*

Proof: We closely follow the description from [DETT99, Section 11.2] and use their notations and definitions. The logic engine consists of the following parts (we will mostly refer to Figures. 4.14 and 4.16 to explain how the parts of the logic engine are connected). The *frame*

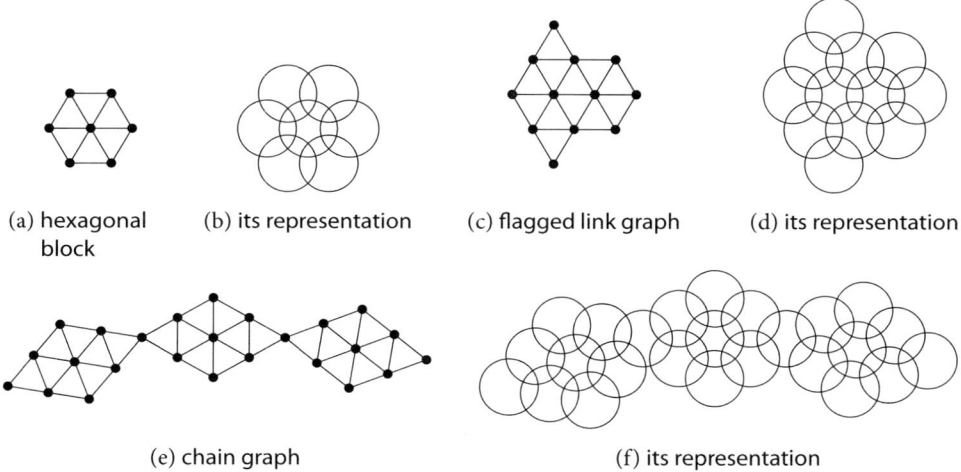

(a) hexagonal
block

(b) its representation

(c) flagged link graph

(d) its representation

(e) chain graph

(f) its representation

Figure 4.15: Gadgets for the logic engine.

and *armatures* (drawn blue and black respectively in Figure 4.14, only half of the drawing is illustrated, the other half is symmetric with respect to the shaft of the logic engine, which is defined below) for the logic graph are built of hexagonal blocks, as shown in Figure 4.15a whose orthogonal unit circle intersection representation is shown in Figure 4.15b. It is easy to see that they are uniquely drawable (up to rotation, reflection, and translation) since K_3 has a unique orthogonal unit circle intersection representation. Each armature corresponds to a variable in U.

A *chain* graph (represented by gray circles in Figure 4.14) is a sequence of *links*, as shown in Figure 4.15e whose orthogonal unit circle intersection representation is shown in Figure 4.15f. The number of links in a chain corresponds to the number of clauses in C. The *shaft* (green in Figure 4.14) is a simple path and serves as an axle for the armatures, that is, the armatures can be flipped around the shaft. Each armature corresponding to a variable x_j has two chains a_j and $\bar{a}_j$ each suspended between one of the ends of the armature and the shaft. For that reason in an orthogonal unit circle intersection representation each chain is taut.

So far we have described the *universal* part of the logic engine, that is, the part that only depends on the number of clauses in C and the number of variables in U; it is illustrated in Figure 4.14. The frame, armatures, and chain graphs have a unique orthogonal unit circle intersection representation up to flipping armatures (see Figure 4.14), since they are built up of hexagonal blocks which are uniquely drawable. We still need to show that the shaft is taut. This is enforced by the bottom part of the frame. Consider the middle horizontal sequence of circles in the bottom part of the frame that spans the frame from the left side to the right; in light blue in Figure 4.14. It is easy to see that the shaft must be drawn as this sequence, because it consists of the same number of circles and must also span the frame from the left side to the right. Since the sequence is taut, the shaft is also taut. Notice that there is still the freedom of flipping each armature together with its chains around the shaft, that is, it can take two possible positions where one part of the armature is either above or below the shaft. This is the flexibility that allows our logic engine to encode a solution of a NAE3SAT instance.

Figure 4.16: Orthogonal unit circle representation of a customized logic engine; only half of the drawing is present. The neighboring flagged links demarcated by the dashed rectangle collide if and only if they are flipped so that they point towards each other; see Figure 4.17.

Now let us show how to customize the logic engine according to an instance of NAE3SAT. A chain link graph can be extended to a *flagged link* by the addition of three new vertices as shown in Figure 4.15c whose orthogonal unit circle representation is shown in Figure 4.15d. Note that it also has a unique drawing. To simulate the given NAE3SAT instance we replace link graphs with flagged link graphs according to the incidence between literals and clauses. If the literal $x_j \in U$ appears in clause $c_i \in C$, then link i of chain a_j is unflagged. If the literal $\bar{x}_j \in U$ appears in clause $c_i \in C$, then link i of of chain $\bar{a}_j$ is unflagged. For an example see Figure 4.16.

It is easy to see that by adjusting the sizes of the frame and the armatures we can ensure that in an orthogonal unit circle intersection representation of the logic engine two flagged links which lie in the same row and are attached to chains of adjacent armatures *collide* if and only if they are flipped so that they point towards each other; see Figure 4.17. Similarly we can ensure that any flag attached to the chain of the outermost armature collides with the frame if it points toward the front edge of the frame, and any flag attached to the chain of the innermost armature collides with that armature if it points toward the rear. Therefore, we can use [DETT99, Theorem 11.2] to show that the corresponding customized logic engine has an

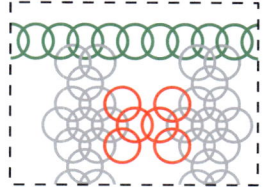

Figure 4.17: The neighboring flagged links collide if and only if they are flipped so that they point towards each other.

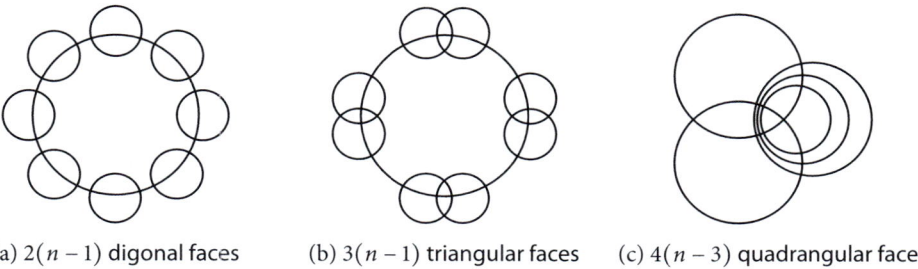

(a) $2(n-1)$ digonal faces   (b) $3(n-1)$ triangular faces   (c) $4(n-3)$ quadrangular faces

Figure 4.18: Arrangements of n orthogonal circles with many digonal, triangular, and quadrangular faces.

orthogonal unit circle representation if and only if the corresponding instance of NAE3SAT is a yes-instance.   □

4.6   Discussions and Open Problems

In Section 4.2 we have provided upper bounds for the number of faces of an orthogonal circle arrangement. As for lower bounds on the number of faces, we found arrangements with n circles containing $2n - 2$ digonal, $3n - 3$ triangular, and $4(n - 3)$ quadrangular faces; see Figures 4.18a, 4.18b, and 4.18c, respectively. Note that the number of digonal faces in the arrangement in Figure 4.18a matches the upper bound in Theorem 4.2. Can we construct better lower bound examples or improve the upper bounds?

Recognizing (unit) disk intersection graphs is $\exists\mathbb{R}$-complete [KM12]. But what is the complexity of recognizing (general) orthogonal circle intersection graphs?

Drawing Graphs with Circular Arcs and Right-Angle Crossings

Once equipped with properties of orthogonal circles from Chapter 4 we can go on to study drawing graphs with circular arcs and right-angle crossings.

A prominent beyond-planar graph class that concerns the crossing angles is the class of k-bend right-angle-crossing graphs [DEL11], or RAC_k graphs for short, that admit a drawing where all crossings form $90°$ angles and each edge is a poly-line with at most k bends. Using right-angle crossings and few bends is motivated by several cognitive studies suggesting a positive correlation between large crossing angles or small curve complexity and the readability of a graph drawing [Hua07, HEH14, HHE08]. Didimo et al. [DEL11] studied the edge density of RAC_k graphs. They showed that RAC_0 graphs with n vertices have at most $4n - 10$ edges (which is tight), that RAC_1 graphs have at most $O(n^{4/3})$ edges, that RAC_2 graphs have at most $O(n^{7/4})$ edges and that all graphs are RAC_3. Dujmović et al. [DGMW10] gave an alternative simple proof of the $4n - 10$ bound for RAC_0 graphs using charging arguments similar to those of Ackerman and Tardos [AT07] and Ackerman [Ack09]. Arikushi et al. [AFK$^+$12] improved the upper bounds to $6.5n - 13$ for RAC_1 graphs and to $74.2n$ for RAC_2 graphs. The bound of $6.5n - 13$ for RAC_1 graphs was also obtained by charging arguments. They also provided a RAC_1 graph with $4.5n - O(\sqrt{n})$ edges. The best known lower and upper bound for the maximum edge density of RAC_1 graphs of $5n - 10$ and $5.5n - 11$, respectively, are due to Angelini et al. [ABFK18].

We extend the class of RAC_0 graphs by allowing edges to be drawn as circular arcs but still requiring $90°$ crossings.

Two circular arcs α and β are orthogonal if they intersect and the underlying circles (that contain the arcs) are orthogonal. For the remainder of this chapter, all arcs will be circular arcs. We consider any straight-line segment to be an arc with infinite radius. Note, though, that Observations 4.1 and 4.2 do not hold for (pairs of) circles of infinite radius. As in the case of circles, for any arc γ of finite radius, let $C(\gamma)$ be its center.

We call a drawing of a graph an *arc-RAC drawing* if the edges are drawn as arcs and any pair of intersecting arcs is orthogonal; see Figure 5.1. A graph that admits an arc-RAC drawings is called an *arc-RAC graph*.

An immediate restriction on the edge density of arc-RAC graphs follows from the fact that there are no four pairwise orthogonal circles; see Lemma 4.1.

Lemma 5.1. *In an arc-RAC drawing, there cannot be four pairwise orthogonal arcs.*

It follows from Lemma 5.1 that arc-RAC graphs are *4-quasi-planar*, that is, an arc-RAC drawing cannot have four edges that pairwise cross. This implies that an arc-RAC graph with n vertices can have at most $72(n - 2)$ edges [Ack09].

Our main contribution is that we reduce this bound to $14n - 12$ using charging arguments similar to those of Ackerman [Ack09] and Dujmović et al. [DGMW10]; see Section 5.1. For us, the main challenge was to apply these charging arguments to a modification of an arc-RAC

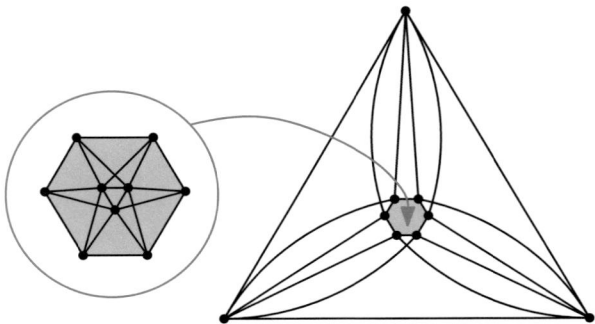

Figure 5.1: An arc-RAC drawing of a graph; this graph is not RAC_0 [BBH+17].

drawing and to exploit, at the same time, geometric properties of the original arc-RAC drawing to derive the bound. We also provide a lower bound of $4.5n - O(\sqrt{n})$ on the maximum edge density of arc-RAC graphs based on the construction of Arikushi et al. [AFK+12]; see Section 5.2. We conclude with some open problems in Section 5.3.

As usual, we forbid vertices to lie in the relative interior of an edge and we do not allow edges to *touch*, that is, to have a common point in their relative interiors without crossing each other at this point. Hence an *intersection point* of two edges is always a *crossing*. When we say that two edges *share a point*, we mean that they either cross each other or have a common endpoint.

5.1   An Upper Bound for the Maximum Edge Density

Let G be a 4-quasi-planar graph, and let D be a 4-quasi-planar drawing of G. In his proof of the upper bound on the edge density of 4-quasi-planar graphs, Ackerman [Ack09] first modified the given drawing so as to remove faces of small degree. We use a similar modification that we now describe.

Consider two edges e_1 and e_2 in D that intersect multiple times. A region in D bounded by pieces of e_1 and e_2 that connect two consecutive crossings or a crossing and a vertex of G is called a *lens*. If a lens is adjacent to a crossing and a vertex of G, then we call such a lens a *1-lens*, otherwise a *0-lens*. A lens that does not contain a vertex of G is *empty*. Every drawing with 0-lenses has a *smallest* empty 0-lens, that is, an empty 0-lens that does not contain any other empty 0-lenses in its interior. We can swap [PRT06, Ack09] the two curves that bound a smallest empty 0-lens; see Figure 5.2. We call such a swap a *simplification step*. Since a simplification step resolves a smallest empty lens, we observe the following.

Observation 5.1. *A simplification step does not introduce any new pairs of crossing edges or any new empty lenses.*

We exhaustively apply simplification steps to our drawing and refer to this as the *simplification process*. Observation 5.1 guarantees that applying the *simplification process* to a drawing D terminates, that is, it results in an empty-0-lens-free drawing D' of G. We call

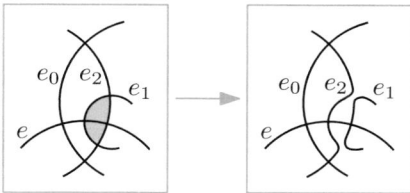

Figure 5.2: A simplification step resolves a smallest empty o-lens; if two edges e_1 and e_2 change the order in which they cross the edge e, they form an empty o-lens intersecting e before the step, and thus, in the original 4-quasi-planar drawing.

the resulting drawing D' *simplified*; it is a *simplification* of D. Observation 5.1 implies the following important property of any simplification step.

Observation 5.2. *Applying a simplification step to a 4-quasi-planar drawing yields a 4-quasi-planar drawing.*

As mentioned above, Ackerman [Ack09] used a similar modification to prepare a 4-quasi-planar drawing for his charging arguments; note, that unlike Ackerman, we do not resolve 1-lenses. We look at the simplification process in more detail, in particular, we consider how it changes the order in which edges cross.

Lemma 5.2. *Let D be an arc-RAC drawing, and let D' be a simplification of D. If two edges e_1 and e_2 cross another edge e in D' in an order different from that in D, then e_1 and e_2 form an empty 0-lens intersecting e in D.*

Proof: Let e_1 and e_2 be two edges as in the statement of the lemma. Then there is a simplification step i where the order in which e_1 and e_2 cross e changes. Let D_i be the drawing immediately before simplification step i, and let D_{i+1} be the drawing right after step i. By construction, the order in which e_1 and e_2 cross e is different in D_i and in D_{i+1}. Since D_i is 4-quasi-planar (see Observation 5.2) and since we always resolve a smallest empty 0-lens, the edges e_1 and e_2 form a smallest empty 0-lens in D_i; see Figure 5.2. Given that the simplification process does not introduce new empty lenses (see Observation 5.1), e_1 and e_2 form an empty 0-lens in the original 4-quasi-planar drawing. $\qquad\square$

We now focus on the special type of 4-quasi-planar drawings we are interested in. Suppose that G is an arc-RAC graph, D is an arc-RAC drawing of G, and D' is a simplification of D. Note that, in general, D' is not an arc-RAC drawing. If two edges e_1 and e_2 cross in D', then they do not form an empty 0-lens in D. This holds because for any two edges forming an empty 0-lens in D, the simplification process removes both of their crossings; therefore, in D' the two edges do not have any crossings. If e_1 and e_2 are incident to the same vertex, they also do not form an empty 0-lens in D, as otherwise they would share three points in D (the two crossing points of the lens and the common vertex of G). Thus, we have the following observation.

Observation 5.3. *Let D be an arc-RAC drawing, and let D' be a simplification of D. If two edges e_1 and e_2 share a point in D', then they do not form an empty 0-lens in D.*

In the following, we first state the main theorem of this section and provide the structure of its proof (deferring one small lemma and the main technical lemma until later). Then, we prove the remaining technical details in Lemmas 5.3 to 5.7 to establish the result.

Theorem 5.1. *An arc-RAC graph with n vertices can have at most* $14n - 12$ *edges.*

Proof: Let G be an arc-RAC graph with a vertex set V and an edge set E, let D be an arc-RAC drawing of G, let D' be a simplification of D, and let G' be the planarization of D', denote the vertices of G' as V' and the edges of G' as E'. Our charging argument consists of three steps.

First, each face f of G' is assigned an initial charge $ch(f) = |f| + v(f) - 4$, where $|f|$ is the degree of f in the planarization and $v(f)$ is the number of vertices of G on the boundary of f. Applying Euler's formula several times, Ackerman and Tardos [AT07] showed that $\sum_{f \in G'} ch(f) = 4n - 8$, where n is the number of vertices of G. In addition, we set the charge $ch(v)$ of a vertex v of G to $16/3$. Hence the total charge of the system is $4n - 8 + 16n/3 = 28n/3 - 8$.

In the next two steps (described below), similarly to Dujmović et al. [DGMW10], we redistribute the charges among faces of G' and vertices of G so that, for every face f, the final charge $ch_{\text{fin}}(f)$ is at least $v(f)/3$ and the final charge of each vertex is non-negative. Observing that

$$28n/3 - 8 \geq \sum_{f \in G'} ch_{\text{fin}}(f) \geq \sum_{f \in G'} v(f)/3 = \sum_{v \in G} \deg(v)/3 = 2|E|/3$$

yields that the number of edges of G is at most $14n - 12$ as claimed. (The second-last equality holds since both sides count the number of vertex–face incidences in G'.)

After the first charging step above, it is easy to see that $ch(f) \geq v(f)/3$ holds if $|f| \geq 4$. We call a face f of G' a k-triangle, k-quadrilateral, or k-pentagon if f has the corresponding shape and $v(f) = k$. Similarly, we call a face of degree two a *digon*. Note that any digon is a 1-digon since all empty 0-lenses have been simplified.

After the first charging step, each digon and each 0-triangle has a charge of -1, and each 1-triangle has a charge of 0. Thus, in the second charging step, we need to find $4/3$ units of charge for each digon, one unit of charge for each 0-triangle, and $1/3$ unit of charge for each 1-triangle. Note that all other faces including 2- and 3-triangles already have sufficient charge.

To charge a digon d incident to a vertex v of G, we decrease $ch(v)$ by $4/3$ and increase $ch(d)$ by $4/3$; see Figure 5.3a. We say that v *contributes* charge to d.

To charge triangles, we proceed similarly to Ackerman [Ack09] and Dujmović et al. [DGMW10, Theorem 7].

Consider a 1-triangle t_1. Let v be the unique vertex incident to t_1, and let $s_1 \in E'$ be the edge of t_1 opposite of v; see Figure 5.3b. Note that the endpoints of s_1 are intersection points in D'. Let f_1 be the face on the other side of s_1. If f_1 is a 0-quadrilateral, then we consider its edge $s_2 \in E'$ opposite to s_1 and the face f_2 on the other side of s_2. We continue iteratively until we meet a face f_k that is not a 0-quadrilateral. If f_k is a triangle, then all the faces $t_1, f_1, f_2, \ldots, f_k$ belong to the same empty 1-lens l incident to the vertex v of t_1. In this case, we decrease $ch(v)$ by $1/3$ and increase $ch(t_1)$ by $1/3$; see Figure 5.3a. Otherwise, f_k is not a triangle and $|f_k| + v(f_k) - 4 \geq 1$ (see Figure 5.3b). In this case, we decrease $ch(f_k)$ by $1/3$ and increase $ch(t_1)$ by $1/3$. We say that the face f_k contributes charge to the triangle t_1 *over its side* s_k.

For a 0-triangle t_0, we repeat the above charging over each side. If the last face on our path is a triangle t', then t_0 and t' are contained in an empty 1-lens (recall that D' does not contain

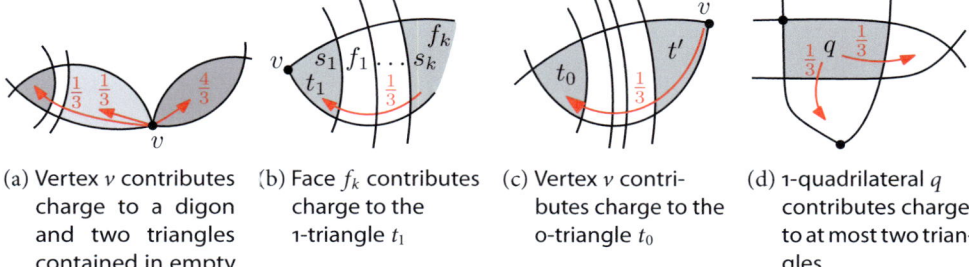

(a) Vertex v contributes charge to a digon and two triangles contained in empty 1-lenses

(b) Face f_k contributes charge to the 1-triangle t_1

(c) Vertex v contributes charge to the 0-triangle t_0

(d) 1-quadrilateral q contributes charge to at most two triangles

Figure 5.3: Transferring charge from vertices and high-degree faces to small-degree faces.

empty 0-lenses) and t' is a 1-triangle incident to a vertex v of G. In this case, we decrease $ch(v)$ by 1/3 and increase $ch(t_0)$ by 1/3; see Figure 5.3c.

Thus, at the end of the second step, the charge of each digon and triangle f is at least $v(f)/3$. Note that the charge of f comes either from a higher-degree face or from a vertex v incident to an empty 1-lens containing f.

In the third step, we do not modify the charging any more, but we need to ensure that

(i) $ch(f) \geq v(f)/3$ still holds for each face f of G' with $|f| \geq 4$ and

(i) $ch(v) \geq 0$ for each v of G.

We first show statement (i). Ackerman [Ack09] noted that a face f with $|f| \geq 4$ can contribute charges over each of its edges at most once. Moreover, f can contribute at most one third unit of charge over each of its edges. Therefore, if $|f| + v(f) \geq 6$, then in the worst case (that is, f contributes charge over each of its edges) f still has a charge of $|f| + v(f) - 4 - |f|/3 \geq v(f)/3$. Thus, it remains to verify that 1-quadrilaterals and 0-pentagons, which initially had only one unit of charge, have a charge of at least 1/3 unit or zero, respectively, at the end of the second step.

A 1-quadrilateral q can contribute charge to at most two triangles since the endpoints of any edge of G' over which a face contributes charge must be intersection points in D'; see Figure 5.3d and recall that q now plays the role of f_k in Figure 5.3b.

A 0-pentagon cannot contribute charge to more than three triangles; see Lemma 5.7.

Now we show statement (ii). Recall that a vertex v can contribute charge to a digon incident to v or to at most two triangles contained in an empty 1-lens incident to v. Observe that two empty 1-lenses with either triangles or a digon taking charge from v cannot overlap; see Figure 5.3a. We show in Lemma 5.3 that v cannot be incident to more than four such empty 1-lenses. In the worst case, v contributes 4/3 units of charge to each of the at most four incident digons representing these empty 1-lenses. Thus, v has non-negative charge at the end of the second step. □

Lemma 5.3. *In any simplified arc-RAC drawing, each vertex is incident to at most four non-overlapping empty 1-lenses.*

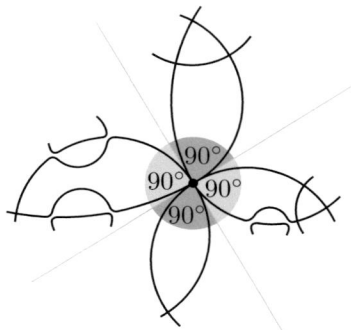

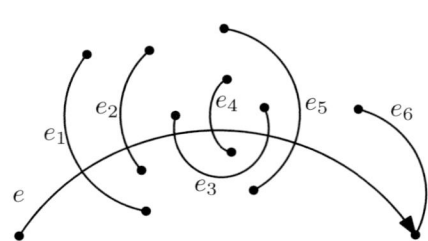

Figure 5.4: The edges of an empty 1-lens form a $\pi/2$ angle at the vertex of the lens.

Figure 5.5: $\Pi(\cdot\,;\cdot)$ describes some intersections along an edge. Here, e.g., $\Pi(e; e_1, e_2, e_3, e_4, e_3, e_5, e_6)$ and $\Pi(e; e_1, e_3, e_4, e_5)$ both hold.

Proof: Let v be a vertex incident to some non-overlapping empty 1-lenses. Consider a small neighborhood of the vertex v in the simplified drawing and notice that in this neighborhood the simplified drawing is the same as the original arc-RAC drawing. Let l be one of the non-overlapping empty 1-lenses incident to v. Then l forms an angle of $90°$ between the two edges incident to v that form l; see Figure 5.4. This is due to the fact that the other "endpoint" of l is an intersection point where the two edges must meet at $90°$. Thus v is incident to at most four non-overlapping empty 1-lenses. □

We now set the stage for proving Lemma 5.7, which shows that a 0-pentagon in a simplified drawing does not contribute charge to more than three triangles. The proof goes by a contradiction. Consider a 0-pentagon that contributes charge to at least four triangles in the simplified drawing. First, we examine which edges of this 0-pentagon cross; see Lemma 5.4. We then describe the order in which these edges share points in the simplified drawing and show that the original arc-RAC drawing must adhere to the same order; see Lemma 5.5. Finally, we use geometric arguments to show that, under these order constraints, an arc-RAC drawing of the edges does not exist; see Lemma 5.6.

Let D be an arc-RAC drawing of some arc-RAC graph G with $V = V(G)$ and $E = E(G)$, let D' be its simplification, and let p be a 0-pentagon that contributes charge to at least four triangles. Let $s_0, s_1, \ldots, s_4$ be the sides of p in clockwise order and denote the edges of G that contain these sides as $e_0, e_1, \ldots, e_4$ so that edge e_0 contains side s_0 etc. Since p contributes charge over at least four sides, these sides are consecutive around p. Without loss of generality, we assume that s_4 is the side over which p does not necessarily contribute charge.

For $i \in \{0, 1, 2, 3\}$, let t_i be the triangle that gets charge from p over the side s_i. The triangle t_i is bounded by the edges e_{i-1} and e_{i+1}. (Indices are taken modulo 5.) Note that all faces bounded by e_{i-1} and e_{i+1} that are between t_i and p must be 0-quadrilaterals. If t_i is a 1-triangle, then e_{i-1} and e_{i+1} are incident to the same vertex of the triangle. Otherwise, t_i is a 0-triangle and e_{i-1} and e_{i+1} cross at a vertex of the triangle. Let $A'_{i-1,i+1}$ denote this common point of e_{i-1} and e_{i+1}, and let $E_p = \{e_0, \ldots, e_4\}$; see Figure 5.6a.

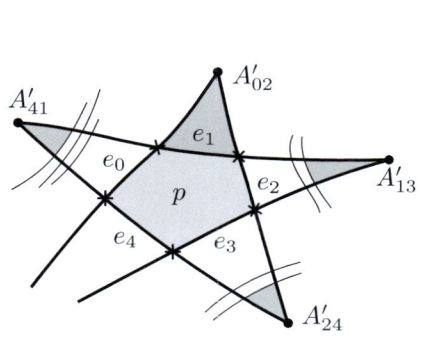

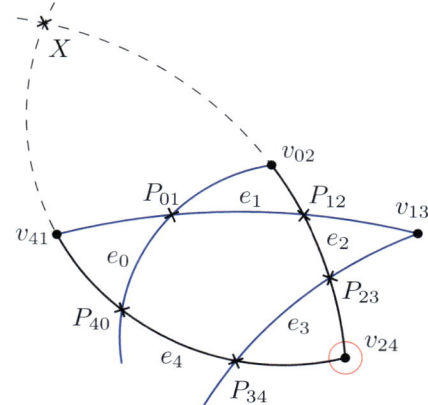

(a) Notation: A o-pentagon p in D' and the edges in E_p. The points of type $A'_{i-1,i+1}$ are either intersection points or vertices of G

(b) Also in D, it holds that $\Pi(e_0; e_4, e_1, e_2)$, $\Pi(e_3; e_1, e_2, e_4)$, and, for $i \in \{1, 2, 4\}$, $\Pi(e_i; e_{i-2}, e_{i-1}, e_{i+1}, e_{i+2})$

Figure 5.6: A 0-pentagon cannot contribute charge to more than three triangles.

We now describe the order in which the edges in E_p share points in D'. To this end, we orient the edges in E_p so that this orientation conforms with the orientation of a clockwise walk around the boundary of p in D'. In addition, we write $\Pi(e_k; e_{i_1}, e_{i_2}, \ldots, e_{i_l})$ if the edge e_k shares points (either crossing points or vertices of the graph) with the edges $e_{i_1}, e_{i_2}, \ldots, e_{i_l}$ in this order with respect to the orientation of e_k; see Figure 5.5. (Note that we can have $\Pi(e_k; e_i, e_j, e_i)$ as edges may intersect twice. We will not consider more than two edges sharing the same endpoint.) Due to the order in which we numbered the edges in E_p, it holds in D' that $\Pi(e_0; e_4, e_1, e_2)$, $\Pi(e_3; e_1, e_2, e_4)$, and, for $i \in \{1, 2, 4\}$, $\Pi(e_i; e_{i-2}, e_{i-1}, e_{i+1}, e_{i+2})$; see Figure 5.6a. Now we show that in D the order is the same. Obviously every pair of edges (e_{i-1}, e_{i+1}) that shares an endpoint in D' also shares an endpoint in D. Furthermore, every pair (e_i, e_{i+1}) or (e_{i-1}, e_{i+1}) of crossing edges crosses in D, too, because the simplification process does not introduce new pairs of crossing edges; see Observation 5.1.

Lemma 5.4. *In the drawing D, the edges e_0 and e_3 do not cross.*

Proof: Assume that the edges e_0 and e_3 cross in D and notice that each of the pairs of edges (e_0, e_1), (e_1, e_2), and (e_2, e_3) forms a crossing in D' (see Figure 5.6a), and hence in D, too. For any arc e, let $\bar{e}$ denote the circle containing e. Recall that a family of *Apollonian circles* [Ogi69, CFKW19] consists of two sets of circles such that each circle in one set is orthogonal to each circle in the other set. Thus, the pairs of circles $(\bar{e}_1, \bar{e}_3)$ and $(\bar{e}_0, \bar{e}_2)$ belong to such a family; the pair $(\bar{e}_1, \bar{e}_3)$ belongs to one set of the family and $(\bar{e}_0, \bar{e}_2)$ belongs to the other set. If the family does not consist of two parabolic pencils, that is, not all of the circles in the family share the same point, which is the case for the circles $\bar{e}_0$, $\bar{e}_1$, $\bar{e}_2$, and $\bar{e}_3$, then one such set consists of hyperbolic pencil, that is, disjoint circles. So either the pair $(\bar{e}_0, \bar{e}_2)$ or the pair

$(\bar{e}_1, \bar{e}_3)$ must be in the hyperbolic pencil. This is a contradiction because each of the two pairs shares a point in D' (see Figure 5.6a), and thus, in D. $\qquad\square$

Lemma 5.5. *In the drawing D, it holds that $\Pi(e_0; e_4, e_1, e_2)$, $\Pi(e_3; e_1, e_2, e_4)$, and, for each $i \in \{1, 2, 4\}$, $\Pi(e_i; e_{i-2}, e_{i-1}, e_{i+1}, e_{i+2})$.*

Proof: Recall that in the drawing D', it holds that $\Pi(e_0; e_4, e_1, e_2)$, $\Pi(e_3; e_1, e_2, e_4)$, and, for each $i \in \{1, 2, 4\}$, $\Pi(e_i; e_{i-2}, e_{i-1}, e_{i+1}, e_{i+2})$; see Figure 5.6a. Consider distinct indices $i, j, k \in \{0, 1, 2, 3, 4\}$ so that the edges e_i and e_j share points with e_k in this order in D', that is, $\Pi(e_k; e_i, e_j)$ in D'. We will show that the edges e_i and e_j share points with e_k in the same order in D, that is, $\Pi(e_k; e_i, e_j)$ in D. In other words, the order in which the edges in E_p share points in D is the same as in D'.

First, note that if the edge e_i or the edge e_j shares an endpoint with e_k, then e_i and e_j do not change the order in which they share points with e_k. This is due to the fact that the simplification process does not modify the graph. Therefore, e_i and e_j share points with e_k in the same order in D as in D', that is, $\Pi(e_k; e_i, e_j)$ in D.

Assume now that both e_i and e_j cross e_k.

If $(i, j) \in \{(0, 3), (3, 0)\}$, then, according to Lemma 5.4, the edges e_i and e_j do not cross in D, so they do not form an empty 0-lens in D, and thus, by Lemma 5.2, e_i and e_j cross e_k in the same order in D as in D', that is, $\Pi(e_k; e_i, e_j)$ in D.

Otherwise, the edges e_i and e_j share a point in D'; see Figure 5.6a. Therefore, by Observation 5.3, e_i and e_j do not form an empty 0-lens in D, and thus, by Lemma 5.2, e_i and e_j cross e_k in the same order in D as in D', that is, $\Pi(e_k; e_i, e_j)$ in D. $\qquad\square$

Thus, we have shown that the order in which the edges in E_p share points in D is the same as in D', see Figure 5.6b. We show now that an arc-RAC drawing with this order does not exist; see Lemma 5.6. This is the main ingredient to prove Lemma 5.7, which says that a 0-pentagon in a simplified arc-RAC drawing contributes charge to at most three triangles.

For simplicity of presentation and without loss of generality, we assume that the points $A'_{i-1,i+1}$ are vertices of G, which we denote by $v_{i-1,i+1}$.

Lemma 5.6. *The edges in E_p do not admit an arc-RAC drawing where it holds that $\Pi(e_0; e_4, e_1, e_2)$, $\Pi(e_3; e_1, e_2, e_4)$, and, for $i \in \{1, 2, 4\}$, $\Pi(e_i; e_{i-2}, e_{i-1}, e_{i+1}, e_{i+2})$.*

Proof: Assume that the edges in E_p admit an arc-RAC drawing where they share points in the order indicated above. For $i \in \{0, \dots, 4\}$, let $P_{i,i+1}$ be the intersection point of e_i and e_{i+1}; see Figure 5.6b. Note that on e_i, the point $P_{i-1,i}$ is before the point $P_{i,i+1}$ (due to $\Pi(e_i; e_{i-1}, e_{i+1})$).

Recall that an *inversion* [Ogi69] with respect to a circle α, the *inversion circle*, is a mapping that takes any point $P \neq C(\alpha)$ to a point P' on the straight-line ray from $C(\alpha)$ through P so that $|C(\alpha)P'| \cdot |C(\alpha)P| = r(\alpha)^2$. Inversion maps each circle not passing through $C(\alpha)$ to another circle and each circle passing through $C(\alpha)$ to a line. The center of the inversion circle is mapped to the "point at infinity". It is known that inversion preserves angles.

We invert the drawing of the edges in E_p with respect to a small inversion circle centered at v_{24}. Let e_i° be the image of e_i, $v_{i-1,i+1}^\circ$ be the image of $v_{i-1,i+1}$ (v_{24}° is the point at infinity),

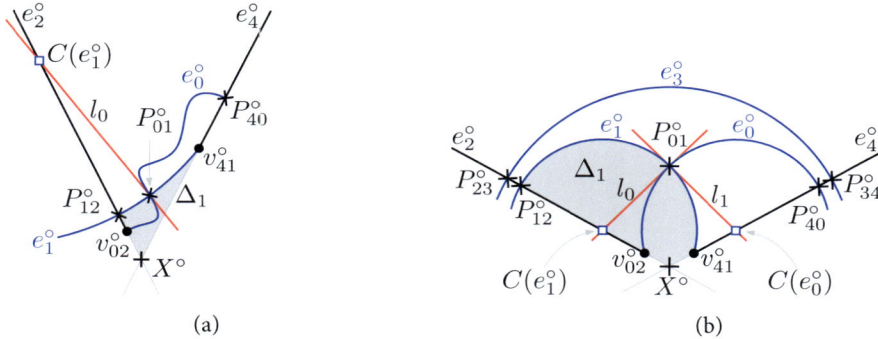

(a)   (b)

Figure 5.7: Illustration for the proof of Lemma 5.6 when e_2 and e_4 belong to two different circles. Image of the inversion with respect to the red circle in Figure 5.6b.

and $P_{i,i+1}^\circ$ be the image of $P_{i,i+1}$. Because in the pre-image the arcs e_2 and e_4 pass through v_{24}, in the image e_2° and e_4° are straight-line rays. We assume that in the image e_2° meets e_4° at the point at infinity, that is, at v_{24}°. Then, taking into account that inversion is a continuous and injective mapping, the order in which the edges in E_p share points is the same in the image.

We consider two cases regarding whether the edges e_2 and e_4 belong to two different circles or not.

Case I: e_2 and e_4 belong to two different circles.

One of the intersection points of their circles is v_{24}, and we let X denote the other intersection point. Here we have that e_2° and e_4° are two straight-line rays meeting at infinity at v_{24}°. Their supporting lines are different and intersect at X°, which is the image of X; see Figure 5.7.

We now assume for a contradiction that the arc e_1° forms a concave side of the triangle $\Delta_1 = P_{12}^\circ v_{41}^\circ X^\circ$; see Figure 5.7a where the triangle is filled gray. (Symmetrically, we can show that the arc e_0° cannot form a concave side of the triangle $\Delta_0 = P_{40}^\circ v_{02}^\circ X^\circ$.) By Observation 4.2, $C(e_1^\circ)$ must lie on the ray e_2°. Since we assume that the arc e_1° forms a concave side of the triangle Δ_1, $C(e_1^\circ)$ and v_{02}° are separated by P_{12}° on e_2°. Consider the tangent l_0 to e_0° at P_{01}°. Again in light of Observation 4.2, l_0 has to go through $C(e_1^\circ)$ because e_0° and e_1° are orthogonal. On the one hand, v_{02}° is to the same side of l_0 as P_{12}°; see Figure 5.7a. On the other hand, l_0 separates P_{12}° and v_{41}° due to $\Pi(e_1; e_4, e_0, e_2)$. Moreover, l_0 does not separate v_{41}° and P_{40}° since it intersects the line of e_4° when leaving the gray triangle Δ_1. So the two points v_{02}° and P_{40}° of the same arc e_0° are separated by l_0, which is a tangent of this arc; contradiction.

Thus, the arc e_1° forms a convex side of the triangle Δ_1, and e_0° forms a convex side of Δ_0; see Figure 5.7b. Now, due to Observation 4.2, $C(e_0^\circ)$ is between v_{41}° and P_{40}°, and $C(e_1^\circ)$ is between v_{02}° and P_{12}°, because that is where the tangents l_1 of e_1° and l_0 of e_0° in P_{01}° intersect the lines of e_4° and e_2°, respectively. Taking into account that $C(e_3^\circ) = X^\circ$, because e_3° is orthogonal to both e_2° and e_4°, we obtain that the points $C(e_3^\circ)$, $C(e_1^\circ)$, P_{12}°, P_{23}° appear on the line of e_2° in this order. Thus, the circle of e_1° is contained within the circle of e_3°. This is a contradiction because e_3° and e_1° must share a point; namely v_{13}°.

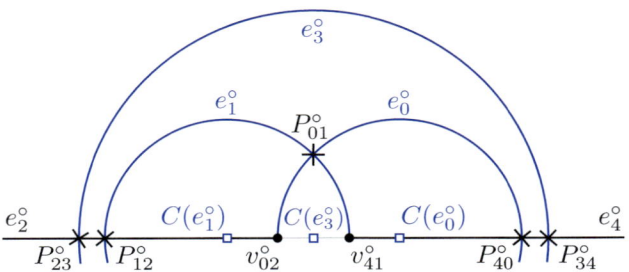

Figure 5.8: Illustration to the proof of Lemma 5.6 when e_2 and e_4 belong to the same circle. Image of the inversion with respect to the red circle in Figure 5.6b.

Case II: e_2 and e_4 belong to the same circle.

Here e_2° and e_4° are two disjoint straight-line rays on the same line l (meeting at infinity at v_{24}°); see Figure 5.8. We direct l as e_4° and e_2° (from right to left in Figure 5.8). Because e_0°, e_1°, and e_3° are orthogonal to l, their centers have to be on l. Due to our initial assumption, we have $\Pi(e_4; e_2, e_3, e_0, e_1)$ and $\Pi(e_2; e_0, e_1, e_3, e_4)$. Hence, along l, we have P_{34}°, P_{40}°, v_{41}°, (on e_4°) and then v_{02}°, P_{12}°, P_{23}° (on e_2°). Therefore, the circle of e_1° is contained in that of e_3°. Hence, e_1° does not share a point with e_3°; a contradiction. □

Lemma 5.7. *A 0-pentagon in a simplified arc-RAC drawing contributes charge to at most three triangles.*

Proof: As discussed above, if a 0-pentagon formed by edges $e_0, e_1, \ldots, e_4$ contributes charge to more than three triangles in a simplified drawing (see Figure 5.6a), then this implies the existence of an arc-RAC drawing where it holds that $\Pi(e_0; e_4, e_1, e_2)$, $\Pi(e_3; e_1, e_2, e_4)$ and, for $i \in \{1, 2, 4\}$, $\Pi(e_i; e_{i-2}, e_{i-1}, e_{i+1}, e_{i+2})$; see Figure 5.6b. This, however, contradicts Lemma 5.6. □

With the proofs of Lemmas 5.3 and 5.7 now in place, the proof of Theorem 5.1 is complete.

5.2   A Lower Bound for the Maximum Edge Density

In this section, we construct a family of arc-RAC graphs with high edge density. Our construction is based on a family of RAC_1 graphs of high edge density that Arikushi et al. [AFK⁺12] constructed. Let G be an embedded graph whose vertices are the vertices of the hexagonal lattice clipped inside a rectangle; see Figure 5.9a. The edges of G are the edges of the lattice and, inside each hexagon that is bounded by the cycle $(P_0, \ldots, P_5)$, six additional edges $(P_i, P_{i+2 \bmod 6})$ for $i \in \{0, 1, \ldots, 5\}$; see Figure 5.9b. We refer to a part of the drawing made up of a single hexagon and its diagonals as a *tile*. In Theorem 5.2 below, we show that each hexagon can be drawn as a regular hexagon and its diagonals can be drawn as two sets of arcs $A = \{\alpha_0, \alpha_1, \alpha_2\}$ and $B = \{\beta_0, \beta_1, \beta_2\}$, so that the arcs in A are pairwise orthogonal, the arcs in B are pairwise non-crossing, and for each arc in B intersecting another arc in A the two

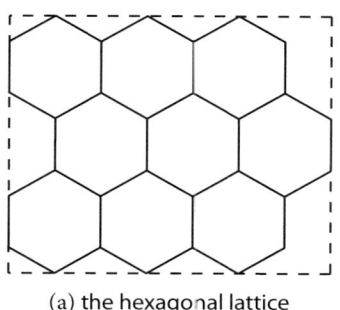

(a) the hexagonal lattice

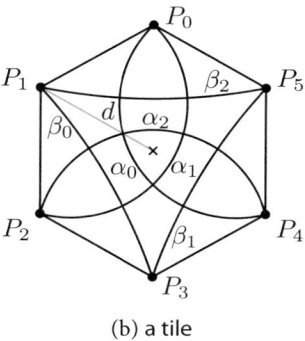

(b) a tile

Figure 5.9: Tiling used for the lower-bound construction.

arcs are orthogonal; we use this construction to establish the theorem. In particular, the arcs in A form the 3-cycle (P_0, P_2, P_4), and the arcs in B form the 3-cycle (P_1, P_3, P_5).

We first define the radii and centers of the arcs in a tile and show that they form only orthogonal crossings. We use the geometric center of the tile as the origin of our coordinate system in the following analysis. We now discuss the arcs in A; then we turn to the arcs in B. For each $j \in \{0, 1, 2\}$, the arc α_j has radius $r_A = 1$ and center $C(\alpha_j) = (d_A \cos(\pi/6 + j\frac{2\pi}{3}), d_A \sin(\pi/6 + j\frac{2\pi}{3}))$, where $d_A = \sqrt{2/3}$ is the distance of the centers from the origin; see Figure 5.10a.

Lemma 5.8. *The arcs in A are pairwise orthogonal.*

Proof: Consider the equilateral triangle $\triangle C(\alpha_0)C(\alpha_1)C(\alpha_2)$ formed by the centers of the three arcs in A. Because the origin is in the center of the triangle, the edge length of the triangle is $2d_A \cos \pi/6 = \sqrt{2}$, and so the distance between the centers of any two arcs is $\sqrt{2}$. The radii of the arcs are 1, hence by Observation 4.1, every two arcs are orthogonal. $\square$

As in Figure 5.10b, for each $j \in \{0, 1, 2\}$, the arc β_j has radius $r_B = \sqrt{\frac{70+40\sqrt{3}}{6}}$ and center $C(\beta_j) = (d_B \cos(\frac{\pi}{2} + \frac{(j+1)2\pi}{3}), d_B \sin(\frac{\pi}{2} + \frac{(j+1)2\pi}{3}))$, where $d_B = \sqrt{\frac{1}{6} + \sqrt{\frac{73+40\sqrt{3}}{6}}}$ is the distance of the centers from the origin.

Lemma 5.9. *If an arc in B intersects an arc in A, then the two arcs are orthogonal.*

Proof: Let $i, j \in \{0, 1, 2\}$. If $j \neq i$, $\|C(\alpha_i) - C(\beta_j)\|^2 = \frac{76+40\sqrt{3}}{6} = 1 + \frac{70+40\sqrt{3}}{6} = r_A^2 + r_B^2$, so by Observation 4.1 α_i and β_j are orthogonal. Otherwise, for $i \in \{0, 1, 2\}$, $\|C(\alpha_i) - C(\beta_i)\| = \sqrt{\frac{112+64\sqrt{3}}{6}} > 1 + \sqrt{\frac{70+40\sqrt{3}}{6}} = r_A + r_B$, so α_i and β_i do not intersect. $\square$

Theorem 5.2. *For infinitely many values of n, there exists an n-vertex arc-RAC graph with $4.5n - O(\sqrt{n})$ edges.*

Proof: We first construct a tile and show that its drawing is indeed a valid arc-RAC drawing. Then it is easy to draw an embedded graph G with the claimed edge density.

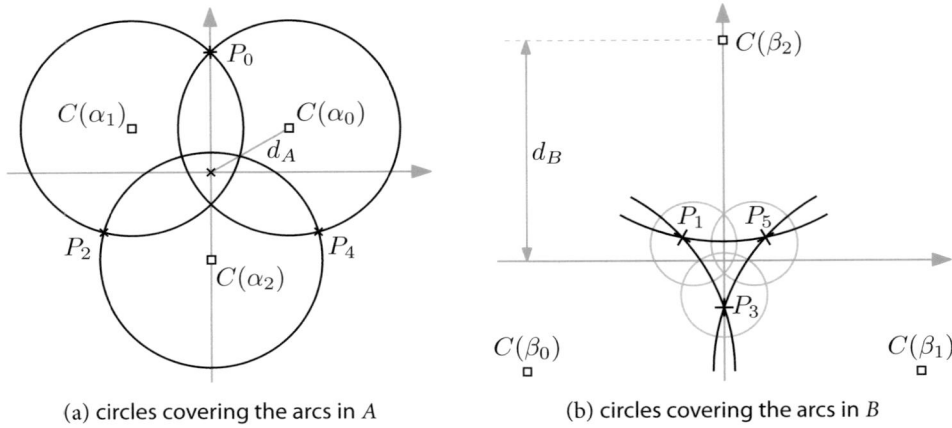

(a) circles covering the arcs in A (b) circles covering the arcs in B

Figure 5.10: Construction for the lower bound on the maximum edge density of arc-RAC graphs.

Consider two circles α and β that intersect in two points of different distance from the origin. Let $X_{\alpha\beta}^-$ be the intersection point that is closer to the origin, and let $X_{\alpha\beta}^+$ be the intersection point further from the origin.

Let the vertices of the hexagon in a tile be $P_0 = X_{\alpha_0\alpha_1}^+$, $P_1 = X_{\beta_2\beta_0}^-$, $P_2 = X_{\alpha_1\alpha_2}^+$, $P_3 = X_{\beta_0\beta_1}^-$, $P_4 = X_{\alpha_2\alpha_0}^+$, and $P_5 = X_{\beta_1\beta_2}^-$. Due to the symmetric definitions of the arcs, the angle between two consecutive vertices of the hexagon is $\pi/3$. Moreover, by a simple computation, we see that for each $j \in \{0, 1, 2\}$ and with $d = \sqrt{1/2} + \sqrt{1/6}$ being the distance of the vertices of the hexagon from the origin, we have:

$$P_{2j} = X_{\alpha_j\alpha_{j+1 \bmod 3}}^+ = \left(d\cos(\tfrac{\pi}{2} + j\tfrac{2\pi}{3}),\; d\sin(\tfrac{\pi}{2} + j\tfrac{2\pi}{3})\right)$$
$$P_{2j+3 \bmod 6} = X_{\beta_j\beta_{j+1 \bmod 3}}^- = \left(d\cos(\tfrac{\pi}{6} + (j+2)\tfrac{2\pi}{3}),\; d\sin(\tfrac{\pi}{6} + (j+2)\tfrac{2\pi}{3})\right).$$

Thus, all the vertices of the hexagon are equidistant from its center, so the hexagon is regular. According to Lemmas 5.8 and 5.9 all crossings of the arcs that belong to the same tile are orthogonal. Now we argue that the arcs in A and B are contained in the regular hexagon. To this end, we show that the arcs do not intersect the relative interior of the edges of the hexagon. To see this, take, for example, the arc α_2, which connects P_2 and P_4. The line segment P_2P_4 is orthogonal to the side P_1P_2 of the hexagon. As the center of α_2 is below P_2P_4, the tangent of α_2 in P_2 enters the interior of the hexagon in P_2. Thus, α_2 does not intersect the relative interior of the edge P_1P_2 (or of any other edge) of the hexagon. Similarly we can show that the arcs in B do not intersect the relative interior of an edge of the hexagon. Therefore, each tile is an arc-RAC drawing, and G is an arc-RAC graph.

Almost all vertices of the lattice with the exception of at most $O(\sqrt{n})$ vertices at the lattice's boundary have degree 9 [AFK$^+$12]. Hence G has $4.5n - O(\sqrt{n})$ edges. $\qquad\square$

As any n-vertex RAC graph has at most $4n - 10$ edges [DEL11], we obtain the following.

Corollary 5.1. *The arc-RAC graphs are a proper superclass of the RAC$_0$ graphs.*

5.3   Discussions and Open Problems

An obvious open problem is to tighten the bounds on the edge density of arc-RAC graphs in Theorems 5.1 and 5.2.

Another immediate question is the relation to RAC_1 graphs, which also extend the class of RAC_0 graphs. This is especially intriguing as the best known lower bound for the maximum edge density of RAC_1 graphs is indeed larger than our lower bound for arc-RAC graphs whereas there may be arc-RAC graphs that are denser than the densest RAC_1 graphs.

The relation between RAC_k graphs and 1-planar graphs is well understood [AFK+12, BBH+17, BDL+17, BDE+16, CLWZ19, EL13]. What about the relation between arc-RAC graphs and 1-planar graphs? In particular, is there a 1-planar graph which is not arc-RAC?

We are also interested in the area required by arc-RAC drawings. Are there arc-RAC graphs that need exponential area to admit an arc-RAC drawing? (A way to measure this off the grid is to consider the ratio between the longest and the shortest edge in a drawing.)

Finally, the complexity of recognizing arc-RAC graphs is open, but likely NP-hard.

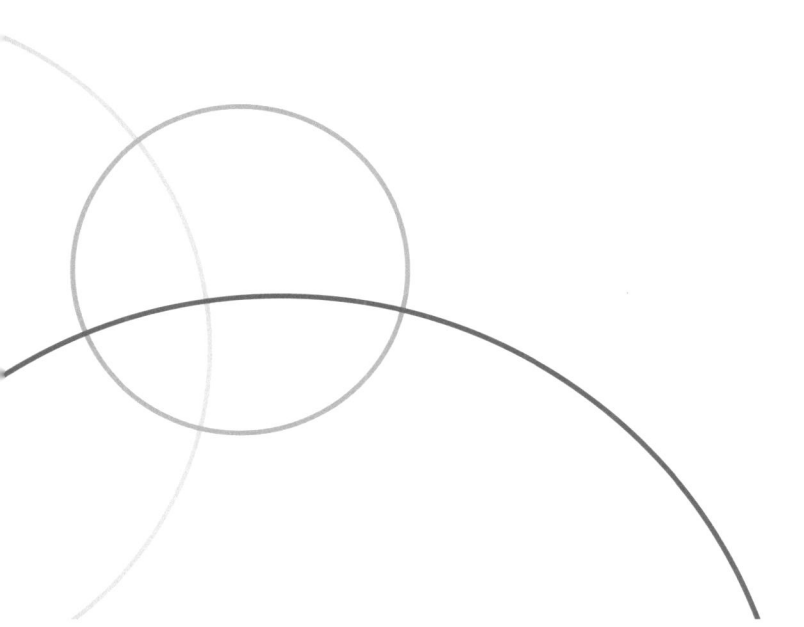

Part II

Optimizing Crossings
in Circular Layouts

Edge Crossing Minimization in Circular Layouts

In contrast to crossing angle optimization in Chapter 5 another way to impose restrictions on crossings is to introduce forbidden patterns on crossing edges in a drawing of a graph. Two commonly studied beyond-planar graph classes with forbidden edge crossing patterns are:

1. *k-planar graphs*, that is, the graphs which can be drawn so that each edge is crossed by at most k other edges.

2. *k-quasi-planar graphs*, that is, the graphs which can be drawn so that no k edges cross pairwise.

Note that the 0-planar graphs and 2-quasi-planar graphs are precisely the planar graphs. Additionally, the 3-quasi-planar graphs are simply called *quasi-planar*.

In this chapter we study these two families of classes of graphs by restricting the drawings to *circular layouts*, that is, so that the vertices are placed in convex position, for example on a circle, and edges routed inside the circle, i.e., we apply the above two generalizations of planar graphs to outerplanar graphs and study *outer k-planarity* and *outer k-quasi-planarity*. For the corresponding graph classes we analyze the following graph parameters (defined in Section 2.1): balanced separators, treewidth, degeneracy, coloring, and edge density. In addition, we consider the recognition problems for these graph classes.

Related work. Ringel [Rin65] was the first to consider k-planar graphs by showing that 1-planar graphs are 7-colorable. This was later improved to 6-colorable by Borodin [Bor84]. This is tight since K_6 is 1-planar. Many additional results on 1-planarity can be found in a recent survey paper [KLM17]. Generally, each n-vertex k-planar graph has at most $3.81n\sqrt{k}$ edges [Ack19] and treewidth $O(\sqrt{kn})$ [DEW17].

Outer k-planar graphs have been considered mostly for $k \in \{0, 1, 2\}$. Of course, the outer 0-planar graphs are the classic outerplanar graphs which are well-known to be 2-degenerate and have treewidth at most 2. It was shown that essentially every graph property is testable on outerplanar graphs [BKN16]. Outer 1-planar graphs are a simple subclass of planar graphs and can be recognized in linear time [ABB+16, HEK+15]. *Full outer 2-planar graphs*, which form a subclass of outer 2-planar graphs, can been recognized in linear time [HN16]. General outer k-planar graphs were considered by Binucci et al. [BGHL18], who showed (among other results) that, for every k, there is a 2-tree which is not outer k-planar. Wood and Telle [WT07] considered a slight generalization of outer k-planar graphs in their work and showed that these graphs have treewidth $O(k)$.

The k-quasi-planar graphs have been heavily studied from the perspective of edge density. The goal here is to settle a conjecture of Pach et al. [PSS96] stating that every n-vertex k-quasi-planar graph has at most $c_k n$ edges, where c_k is a constant depending only on k. This conjecture is true for $k = 3$ [AT07] and $k = 4$ [Ack09]. The best known upper bound is

$(n \log n) 2^{\alpha(n)^{c_k}}$ [FPS13], where α is the inverse of the Ackermann function. Edge density was also considered in the "outer" setting: Capoyleas and Pach [CP92] showed that any outer k-quasi-planar graph with n vertices has at most $2(k-1)n - \binom{2k-1}{2}$ edges. Dress et al. [DKM02] and Nakamigawa [Nak00] showed that there are outer k-quasi-planar graphs meeting this bound (if $n \geq 2k-1$). Actually, the outer k-quasi-planar graphs that meet this bound are exactly the maximal outer k-quasi-planar graphs. (Recall that a graph is *maximal* with respect to a given graph property if adding any edge to the graph destroys the property.) Dress et al. [DKM02] and Nakamigawa [Nak00] showed that for every two maximal outer k-quasi-planar drawings $G = (V, E)$ and $G' = (V, E')$ whose corresponding vertices are in the same positions, there is a sequence of local edge exchange operations (called *flips*) producing drawings $G = G_1, G_2, \ldots, G_t = G'$ such that each intermediate drawing is a maximal k-quasi-planar drawing. More recently, it was shown that the *semi-bar k-visibility graphs* are outer $(k+2)$-quasi-planar [GKT14]. Apart from these results, the outer k-quasi-planar graphs do not seem to have received much attention.

The relationship between k-planar graphs and k-quasi-planar graphs was considered recently. While any k-planar graph is clearly $(k+2)$-quasi-planar, Angelini et al. [ABB$^+$17] showed that any k-planar graph is $(k+1)$-quasi-planar.

The *convex* (or *1-page book*) *crossing number* of a graph [Sch17] is the minimum number of crossings which occur in any convex drawing. This concept has been introduced several times (see [Sch17] for more details). The convex crossing number is NP-complete to compute [MKNF87]. However, recently Bannister and Eppstein [BE18] used treewidth-based techniques (via extended monadic second-order logic or MSO$_2$, see Section 2.4) to show that it can be computed in linear FPT time, i.e., $O(f(c) \cdot n)$ time where c is the convex crossing number and f is a computable function. Thus, for any k, the *outer k-crossing graphs* can be recognized in time linear in $n + m$.

Our contribution.   In Section 6.1, we consider outer k-planar graphs. We show that the largest outer k-planar complete graph has $(\lfloor \sqrt{4k+1} \rfloor + 2)$ vertices. Further we show that each outer k-planar graph is $\lfloor 3.5\sqrt{k} \rfloor$-degenerate. This provides bounds of $\lfloor 3.5\sqrt{k} \rfloor n$ and $\lfloor 3.5\sqrt{k} \rfloor + 1$ on the edge density and the chromatic number of an n-vertex outer k-planar graph respectively. We further show that every outer k-planar graph has separation number at most $2k + 3$. For each fixed k, we use the corresponding balanced separators to obtain a quasi-polynomial time algorithm to test outer k-planarity, i.e., these recognition problems are not NP-hard unless ETH fails.

In Section 6.2, we consider outer k-quasi-planar graphs. We relate outer k-quasi-planar graphs to other graph classes, in particular, planar graphs.

Finally, in Section 6.3, we restrict outer k-planar and outer k-quasi-planar drawings to *full* drawings (where no crossing appears on the boundary), and to *closed* drawings (where the vertex sequence on the boundary is a cycle in the graph). As we have already mentioned, the class of full outer 2-planar graphs have been considered by Hong and Nagamochi [HN16] who showed that full outer 2-planarity testing can be performed in linear time. We first observe that a graph is full outer k-planar (k-quasi-planar) if and only if its maximal biconnected components are closed outer k-planar (k-quasi-planar), this was observed for full outer 2-planar graphs by Hong and Nagamochi [HN16]. Then, for each k, we express both *closed outer k-planarity* and *closed outer k-quasi-planarity* in extended monadic second-order logic (MSO$_2$; see Section 2.4). Thus, since outer *k*-planar graphs have bounded treewidth, full outer *k*-planarity is testable in $O(f(k) \cdot n)$ time, for a computable function f. We note that this result greatly generalizes the work of Hong and Nagamochi [HN16].

6.1   Outer *k*-Planar Graphs

In this section we consider outer *k*-planar graphs. We study their structural properties such as degeneracy and separation number. Based on these structural properties we obtain bounds on the colorability of outer *k*-planar graphs as well as a quasi-polynomial time recognition algorithm.

Degeneracy.   In this section we focus on the degeneracy of outer *k*-planar graphs. First we consider outer *k*-planar complete graphs. We show that the largest outer *k*-planar complete graph has $\lfloor \sqrt{4k+1} \rfloor + 2$ vertices; see Observation 6.1. This implies that there are outer *k*-planar graphs whose minimum degree is $\lfloor \sqrt{4k+1} \rfloor + 1$. We then bound the degeneracy of outer *k*-planar graphs by $\lfloor 3.5\sqrt{k} \rfloor$; see Theorem 6.1.

Observation 6.1.   *For every k, the largest outer k-planar complete graph has at most* $\lfloor \sqrt{4k+1} \rfloor + 2$ *vertices. Moreover, for every k, the complete graph with* $\lfloor \sqrt{4k+1} \rfloor + 2$ *vertices is outer k-planar.*

Proof: Let r be the size of the largest outer *k*-planar complete graph. Consider such a complete graph and its outer *k*-planar drawing. Let e be an edge that splits the complete graph so that there are $(r-2)/2$ many vertices on both sides if r is even, or $(r-1)/2$ on one side and $(r-3)/2$ on the other if r is odd. Then the edge e has the largest number of crossings among all the edges in the drawing, namely $((r-2)/2)^2$ if r is even and $((r-3)(r-1))/4$ if r is odd. Taking into account the fact that no edge is crossed more than k times, we obtain that $r \leq \lfloor \sqrt{4k+1} \rfloor + 2$. Similarly, if $r = \lfloor \sqrt{4k+1} \rfloor + 2$ the drawing is outer *k*-planar.   □

Before we proceed further we introduce some helpful notation. Let G be an outer *k*-planar graph. Consider some outer *k*-planar embedding of G. Without loss of generality we can assume that the vertices are on a circle and the edges are drawn straight-line. We say that an edge ab *splits off* $l \in \mathbb{N}$ vertices of G to one side if one of the open half-planes defined by the edge ab contains exactly l vertices (not including a and b). From the context it will be clear which of the two half-planes we mean.

Theorem 6.1.   *For each k let $\delta^\star$ be the largest minimum degree among all outer k-planar graphs. Then $\delta^\star \leq \lfloor c_k \sqrt{k} \rfloor$ where*

$$c_k = \frac{5}{4\sqrt{k}} + \frac{3}{4}\sqrt{\frac{1}{k} + 8}.$$

The sequence $(c_k)_{k \geq 1}$ is monotonically decreasing with $c_1 = 3.5$ and the limit $3\sqrt{2}/2$.

Proof: Let G be an outer *k*-planar graph whose minimum degree is $\delta^\star$. Consider some outer *k*-planar embedding of G. Assume that there exists an edge e that splits off $t \in \mathbb{N}$ vertices in the embedding of G to one side, then there are at least $\delta^\star t - t(t-1) + 2t = \delta^\star t - t(t+1)$ edges crossing the edge e (on the left-hand side of the equality the second term stands for the sum of the degrees of a clique on t vertices and the third term for the number of edges incident to the endpoints of e and to t many vertices). Because G is outer k-planar, we have that $\delta^\star t - t(t+1) \leq k$. Therefore, either $t \leq t_1$ or $t_2 \leq t$, where $t_1, t_2 = \left((\delta^\star - 1) \pm \sqrt{(\delta^\star - 1)^2 - 4k} \right)/2$, if $\delta^\star \geq 2\sqrt{k} + 1$. Assume for contradiction that $\delta^\star \geq c\sqrt{k}$ for some $c > c_k$ where

$$c_k = \frac{5}{4\sqrt{k}} + \frac{3}{4}\sqrt{\frac{1}{k} + 8}.$$

Then t_1 and t_2 are well defined because $c_k > 1/\sqrt{k} + 2$. Call an edge that splits off at least t_2 vertices to both sides $long$. Observe that each vertex is incident to at least $\delta^\star - 2(t_1 + 1)$ long edges. Then the number l of long edges incident to each vertex is at least $\delta^\star - 2(t_1 + 1) = \delta^\star - \delta^\star - 1 + \sqrt{(\delta^\star - 1)^2 - 4k} \geq \sqrt{(c^2 - 4)k - 2c\sqrt{k} + 1} - 1$. Take a smallest long edge e' in the outer k-planar embedding of G, that is, in one of the open half-planes h defined by e' there is no long edge that is completely contained in h. Let V_h be the set of vertices of the graph that are contained in the half-plane h. Because e' is a long edge, $|V_h| \geq t_2 = \left((\delta^\star - 1) + \sqrt{(\delta^\star - 1)^2 - 4k} \right)/2$. Because it is a smallest long edge all the long edges incident to the vertices in V_h must cross e', therefore, the number of edges that cross e' is at least

$$\ell t_2 \geq \frac{1}{2}\left(\sqrt{(c^2 - 4)k - 2c\sqrt{k} + 1} - 1 \right)\left((c\sqrt{k} - 1) + \sqrt{(c^2 - 4)k - 2c\sqrt{k} + 1} \right).$$

Let

$$f(c, k) = \frac{1}{2}\left(\sqrt{(c^2 - 4)k - 2c\sqrt{k} + 1} - 1 \right)\left((c\sqrt{k} - 1) + \sqrt{(c^2 - 4)k - 2c\sqrt{k} + 1} \right) - k.$$

Consider the equation $f(c, k) = 0$ for any $k \in \mathbb{N}$ and c in the interval $(1/\sqrt{k} + 2, \infty)$. It can be simplified to the following quadratic equation

$$2c^2 k - 5c\sqrt{k} + 2 - 9k = 0. \tag{6.1}$$

For each $k \in \mathbb{N}$ the only root of equation (6.1) with respect to c in the interval $(1/\sqrt{k} + 2, \infty)$ is c_k. Therefore, because $c > c_k$ the number of edges that cross e' is strictly larger than k; contradiction.

Thus, for each $k \in \mathbb{N}$ the largest maximum minimum degree of any outer k-planar graph is at most $c_k\sqrt{k}$. Observe that, $(c_k)_{k \geq 1}$ is a monotonically decreasing sequence with $c_1 = 3.5$ and the limit $3\sqrt{2}/2$. $\square$

It is worth pointing out that the lower bound from Observation 6.1 differs from the upper bound from Theorem 6.1 by at most one for k up to 54. As a direct consequence of Theorem 6.1, we obtain the following.

Corollary 6.1. *Every outer k-planar graph has at most $\lfloor 3.5\sqrt{k} \rfloor n$ edges.*

Note that for, *outer* k-planar graphs, Corollary 6.1 provides a better upper bound than the upper bound [Ack19] of $3.81\sqrt{k}n$ for general n-vertex k-planar graphs ($k \geq 4$).

By combining Observation 6.1 and Theorem 6.1, we obtain the following.

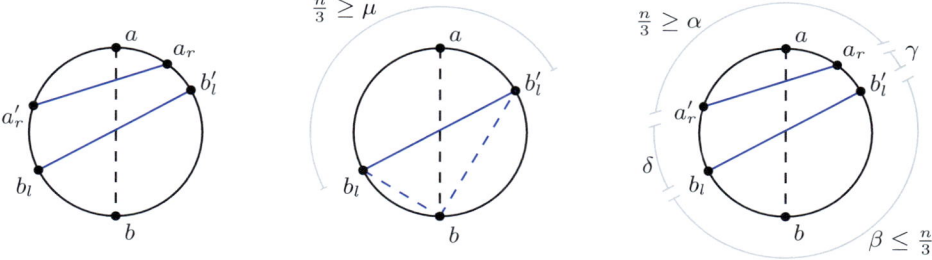

Figure 6.1: Left: the pair of parallel edges $b_l b_l'$ and $a_r a_r'$; center: case 1; right: case 2.

Corollary 6.2. *Every outer k-planar graph can be colored with $\lfloor 3.5\sqrt{k} \rfloor + 1$ colors. There exist outer k-planar graphs that need at least $\lfloor \sqrt{4k+1} \rfloor + 2$ colors.*

Quasi-polynomial time recognition via balanced separators.   We show that outer k-planar graphs have separation number at most $2k+3$ (Theorem 6.2). Via a result of Dvořák and Norin [DN14], this implies they have $O(k)$ treewidth. However, Proposition 8.5 of [WT07] implies that every outer k-planar graph has treewidth at most $3k + 11$, i.e., a better bound on the treewidth than applying the result of Dvořák and Norin to our separators. The treewidth $3k + 11$ bound also implies a separation number of $3k + 12$, but our bound is better. Our separators also allow outer k-planarity testing in quasi-polynomial time (Theorem 6.3).

Theorem 6.2. *Each outer k-planar graph has separation number at most $2k + 3$.*

Proof: Consider an outer k-planar drawing. If the graph has an edge that splits off $[n/3, 2n/3]$ vertices to one side, we can use this edge to obtain a balanced separator of size at most $k + 2$, i.e., by choosing the endpoints of this edge and a vertex cover of the edges crossing it. So, suppose no such edge exists. Consider a pair of vertices (a, b) such that the line ab divides the drawing into left and right sides having an almost equal number of vertices (with a difference at most one). If the edges which cross the line ab also mutually cross each other, there can be at most k of them. Thus, we again have a balanced separator of size at most $k+2$. So, it remains to consider the case when we have a pair of edges that cross the line ab, but do not cross each other. We call such a pair of edges *parallel*. We now pick a pair of parallel edges in a specific way. Starting from b, let b_l be the first vertex along the boundary in clockwise direction such that there is an edge $b_l b_l'$ that crosses the line ab. Symmetrically, starting from a, let a_r be the first vertex along the boundary in clockwise direction such that there is an edge $a_r a_r'$ that crosses the line ab; see Fig. 6.1 (left). Note that the edges $a_r a_r'$ and $b_l b_l'$ are either identical or parallel. In the former case, we see that all other edges crossing the line ab must also cross the edge $a_r a_r' = b_l b_l'$, and as such there are again at most k edges crossing the line ab. In the latter case, there are two subcases that we treat below. For two vertices u and v, let $[u, v]$ be the set of vertices that starts with u and, going clockwise, ends with v. Let $(u, v) = [u, v] \setminus \{u, v\}$.

   Case 1. *The edge $b_l b_l'$ splits off $\mu \le n/3$ vertices to the top; see Fig. 6.1 (center).*
In this case, either $[b_l', b]$ or $[b, b_l]$ has $[n/3, n/2]$ vertices. We claim that neither the line bb_l nor the line bb_l' can be crossed more than k times. Namely, each edge that crosses the line

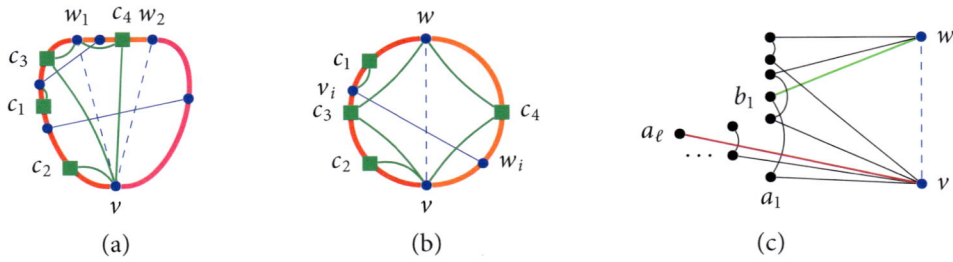

Figure 6.2: Shapes of separators, special separator S in blue, regions in different colors (red, orange, and pink), components connected to blue vertices in green: (a) closest-parallels case; (b) single-edge case; (c) special case for single-edge separators.

bb_l also crosses the edge $b_l b'_l$. Similarly, each edge that crosses the line bb'_l also crosses the edge $b_l b'_l$. Thus, we have a separator of size at most $k + 2$, regardless of whether we choose bb_l or bb'_l to separate the graph. As we observed above, one of them is balanced.

Case 1′. The edge $a_r a'_r$ splits off at most $n/3$ vertices to the bottom.
This is symmetric to case 1.

Case 2. The edge $b_l b'_l$ splits off at most $n/3$ vertices to the bottom, and the edge $a_r a'_r$ splits off at most $n/3$ vertices to the top; see Fig. 6.1 (right).

We show that we can always find a pair of parallel edges such that one splits off at most $n/3$ vertices to the bottom and the other splits off at most $n/3$ vertices to the top, and no edge between them is parallel to either of them. We call such a pair *close*. If there is an edge e between $b_l b'_l$ and $a_r a'_r$, we form a new pair by using e and $a_r a'_r$ if e splits off at most $n/3$ vertices to the bottom or by using e and $b_l b'_l$ if e splits off at most $n/3$ vertices to the top. By repeating this procedure, we always find a close pair. Hence, we can assume that $b_l b'_l$ and $a_r a'_r$ actually form a close pair. Let $\alpha = |(a'_r, a_r)|$, $\beta = |(b'_l, b_l)|$, $\gamma = |(a_r, b'_l)|$, and $\delta = |(b_l, a'_r)|$; see Fig. 6.1 (right).

Suppose that $a'_r = b_l$ or $a_r = b'_l$. We can now use both edges $b_l b'_l$ and $a_r a'_r$ (together with any edges crossing them) to obtain a separator of size at most $2k + 3$. The separator is balanced since $\alpha + \beta \le 2n/3$ and $\gamma + \delta \le 2n/3$.

So, now a_r, a'_r, b_l, b'_l are all distinct. Note that $\gamma, \delta \le n/2$ since each side of the line ab has at most $n/2$ vertices. We separate the graph along the line $b_l a_r$. Namely, all the edges that cross this line must also cross $b_l b'_l$ or $a'_r a_r$. Therefore, we obtain a separator of size at most $2k + 2$.

To see that the separator is balanced, we consider two cases. If $\delta \ge n/3$ (or $\gamma \ge n/3$), then $\alpha + \beta + \gamma \le 2n/3$ (or $\alpha + \beta + \delta \le 2n/3$). Otherwise $\delta < n/3$ and $\gamma < n/3$. In this case $\delta + \alpha \le 2n/3$ and $\gamma + \beta \le 2n/3$. In both cases the separator is balanced. □

Theorem 6.3. *For fixed k, testing the outer k-planarity of an n-vertex graph takes $O(2^{\mathrm{polylog}(n)})$ time.*

Proof: Our approach is to leverage the structure of the balanced separators as described in the proof of Theorem 6.2. Namely, we enumerate the sets which could correspond to such a separator, pick an appropriate outer k-planar drawing of these vertices and their edges, partition the components arising from this separator into *regions*, and recursively test the outer k-planarity of the regions.

To obtain quasi-polynomial runtime, we need to limit the number of components on which we branch. To do so, we group them into regions defined by special edges of the separators.

By the proof of Theorem 6.2, if our input graph has an outer k-planar drawing, there must be a separator which has one of the two shapes depicted in Fig. 6.2 (a) and (b). Here we are not only interested in the up to $2k + 3$ vertices of the balanced separator, but in the set S of up to $4k + 3$ vertices that one obtains by taking both endpoints of the edges used to find the separator. Note: S is also a balanced separator. We use a brute force approach to find such an S. Namely, we first enumerate vertex sets of size up to $4k + 3$. We then consider two possibilities, i.e., whether this set can be drawn similar to one of the two shapes from Fig. 6.2. So, we now fix this set S. Note that since S has $O(k)$ vertices, the subgraph G_S induced by S can have at most a function of k different outer k-planar drawings. Thus, we further fix a particular drawing of G_S.

We now consider the two different shapes separately. In the first case, in S, we have three special vertices v, w_1 and w_2 and in the second case we will have two special vertices v and w. These vertices will be called *boundary* vertices and all other vertices in S will be called *regional* vertices. Note that, since we have a fixed drawing of G_S, the regional vertices are partitioned into regions by the specially chosen boundary vertices. Now, from the structure of the separator which is guaranteed by the proof of Theorem 6.2, no component of $G \setminus S$ can be adjacent to regional vertices which live in different regions with respect to the boundary vertices.

We first discuss the case of using G_S as depicted in Fig. 6.2 (a). Here, we start by picking the three special vertices v, w_1 and w_2 from S to take the role as shown in Fig. 6.2 (a). The following arguments regarding this shape of separator are symmetric with respect to the pair of opposing regions.

Notice that if there is a component connected to regional vertices of different regions, we can reject this configuration. From the proof of Theorem 6.2, we further observe that no component can be adjacent to all three boundary vertices. Namely, this would contradict the closeness of the parallel edges or it would contradict the members of the separator, i.e., it would imply an edge connecting distinct regions. We now consider the four possible different types of components c_1, c_2, c_3 and c_4 in Fig. 6.2 (a) that can occur in a region neighboring w_1. Components of type c_1 are connected to (possibly many) regional vertices of the same region and may be connected to boundary vertices as well. In any valid drawing, they will end up in the same region as their regional vertices. Components of type c_2 are not connected to any regional vertices and only connected to one of the three boundary vertices. Since they are not connected to regional vertices, they can not interfere with other parts of the drawing, so we can arbitrarily assign them to an adjacent region of their boundary vertex. Components that are connected to two boundary vertices appear at first to have two possible placements, e.g., as c_3 or c_4 in Fig. 6.2 (a). However, c_4 is not a valid placement for this type of component since it would contradict the fact that this separator arose from two close parallel edges as argued in the proof of Theorem 6.2. From the above discussion, we see that from a fixed configuration (i.e., set S, drawing of G_S, and triple of boundary vertices), if the drawing of G_S has the shape depicted in Fig. 6.2 (a), we can either reject the current configuration (based on having bad components), or we see that every component of $G \setminus S$ is either attached to exactly one boundary vertex or it has a well-defined placement into the regions defined by the boundary vertices. For those components which are attached to exactly one boundary vertex, we observe that it suffices to recursively produce a drawing of that component together with its boundary vertex and to place this drawing next to the boundary vertex. For the other components, we partition them into their regions and recurse on the regions. This covers all cases for this separator shape.

The other shape of our separator can be seen in Fig. 6.2 (b). Note that we now have two boundary vertices v and w and thus only have two regions. Again we see the two component types c_1 and c_2 and can handle them as above. We also have components connected to both v and w but no regional vertices. These components now truly have two different placement options c_3, c_4. If we have an edge $v_i w_i$ (as in Fig. 6.2 (b)) of the separator that is not vw, we now observe that there cannot be more than k such components. Namely, in any drawing, for each component, there will be an edge connecting this component to either v or w which crosses $v_i w_i$. Thus, we now enumerate all the different placements of these components as type c_3 or c_4 and recurse accordingly.

However, the separator may be exactly the pair (v, w). Note that there are no components of type c_1 and the components of type c_2 can be handled as before. We will now argue that we can have at most a function of k different components of type c_3 or c_4 in a valid drawing. Consider the components of type c_3 (the components of type c_4 can be counted similarly). In a valid drawing, each type c_3 component defines a sub-interval of the left region spanning from its highest to its lowest vertex such that these vertices are adjacent to one of v or w. Two such intervals relate in one of three ways: They overlap, they are disjoint, or one is contained in the other. We group components with either overlapping or disjoint intervals into *layers*. We depict this situation in Fig. 6.2 (c) where, for simplicity, for every component we only draw its highest vertex and its lowest vertex and they are connected by one edge.

Let $a_1 b_1$ be the bottommost component of type c_3 (i.e., a_1 is the clockwise-first vertex from v in a component of type c_3). The first layer is defined as the component $a_1 b_1$ together with every component whose interval either overlaps or is disjoint from the interval of $a_1 b_1$. Now consider the green edge $b_1 w$ (see Fig. 6.2 (c)), note we may have that this edge connects a_1 to w instead. Now, for every component of this layer which is disjoint from the interval of $a_1 b_1$, this edge is crossed by at least one edge connecting it to v. Furthermore, for every component of this layer which overlaps the interval of $a_1 b_1$, there is an edge connecting b_1 to either v or w which is crossed by at least one edge within that component. So in total, there can only be $O(k)$ components in this first layer. New layers are defined by considering components whose intervals are contained in $a_1 b_1$. To limit the total number of layers, let a_ℓ be the bottommost vertex of the first component of the deepest layer and consider the purple edge va_ℓ. This edge is crossed by some edge of every layer above it and as any edge can only have k crossings, there can only be $O(k)$ different levels in total. This leaves us with a total of at most $O(k^2)$ components per region and again we can enumerate their placements and recurse accordingly.

The above algorithm provides the following recurrence regarding its runtime. Let $T(n)$ denote the runtime of our algorithm for an outer k-planar graph with n vertices. Then,

$$T(n) \leq \begin{cases} n^{O(k)} \cdot f(4k+3) \cdot n^3 \cdot n \cdot T(2n/3) & \text{for } n > 5k, \\ f(n) & \text{otherwise,} \end{cases}$$

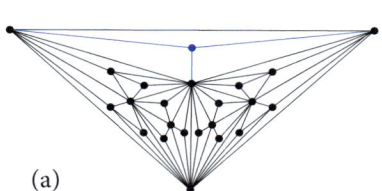 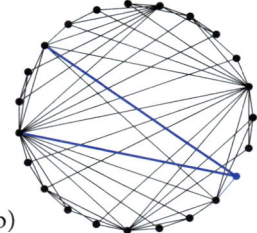

(a)   (b)

Figure 6.3: A vertex-minimal 23-vertex planar 3-tree which is not outer quasi-planar: (a) planar drawing; (b) deleting the blue vertex makes the drawing outer quasi-planar.

where $f(s)$ denotes the number of different outer k-planar drawings of a graph with s vertices. The factor $n^{O(k)}$ stands for finding all possible separators of size $4k + 3$, $f(4k + 3)$ is the number of different outer k-planar drawings of such a separator, n^3 is the time needed to partition the remaining vertices of the graph into regions, n is the largest number of different regions, and $T(2n/3)$ is the runtime of the recursive call on a region.

Thus, the algorithm runs in quasi-polynomial time, i.e., $T \in 2^{\text{polylog}(n)}$. $\qquad\qquad\square$

6.2 Outer k-Quasi-Planar Graphs

In this section we consider outer k-quasi-planar graphs. We first describe some classes of graphs which are outer quasi-planar (outer 3-quasi-planar) and some classes of graphs that are not outer quasi-planar. In particular, we show that there are planar graphs which are not outer quasi-planar what yields the fact that planar graphs and quasi-planar graphs are incomparable; see Theorem 6.4.

Proposition 6.1. *The following graphs are outer 3-quasi-planar: (a) $K_{4,4}$; (b) K_5; (c) planar 3-tree with three complete levels; (d) square-grids of any size.*

Proof: (a) and (b) are easily observed. (c) was experimentally verified by constructing a Boolean expression and using MiniSat to check it for satisfiability; see Section 6.4. (d) follows from square-grids being sub-Hamiltonian. $\qquad\qquad\square$

Correspondingly, we note complete and complete bipartite graphs which are not outer-quasi planar. Furthermore, not all planar graphs are outer quasi-planar, e.g., the vertex-minimal planar 3-tree in Figure 6.3 (a) is not outer quasi-planar, this was verified by checking for satisfiability the corresponding Boolean expression; see Section 6.4. A drawing of the graph in Figure 6.3 (b) was constructed by removing the blue vertex and drawing the remaining graph in an outer quasi-planar way.

Proposition 6.2. *The following graphs are not outer 3-quasi-planar: (a) $K_{p,q}$, $p \geq 3$, $q \geq 5$; (b) K_n, $n \geq 6$; (c) planar 3-trees with at least four complete levels.*

Together, Propositions 6.1 and 6.2 immediately yield the following.

Theorem 6.4. *Planar graphs and outer 3-quasi-planar graphs are incomparable under containment.*

Remark 6.1. For outer k-quasi-planar graphs ($k > 3$) containment questions become more intricate. Every planar graph is outer 5-quasi-planar because planar graphs have page number 4 [Yan89]. We also know a planar graph that is not outer 3-quasi-planar. It is open whether every planar graph is outer 4-quasi-planar.

6.3   Testing for Full Convex Drawings via MSO$_2$

The class of *full outer k-planar graphs* was first introduced by Hong and Nagamochi [HN16]. They are defined as having a convex drawing which is k-planar and additionally there is no crossing on the outer boundary of the drawing. Hong and Nagamochi gave a linear-time recognition algorithm for full outer 2-planar graphs. They state that a graph G is (full) outer 2-planar, if and only if its biconnected components are (full) outer 2-planar and that the outer boundary of a full outer 2-planar embedding of a biconnected graph G is a Hamiltonian cycle of G. We call the subclasses of outer k-planar and outer k-quasi-planar graphs that have a convex drawing where the circular order forms a Hamiltonian cycle *closed outer k-planar* and *closed outer k-quasi-planar* respectively and observe that the property stated by Hong and Nagamochi also carries over to general outer k-planar and outer k-quasi-planar graphs; see Observation 6.2.

Observation 6.2. *A graph G is full outer k-planar (outer k-quasi-planar), if and only if its biconnected components are closed outer k-planar (closed outer k-quasi-planar).*

We show that we can encode closed outer k-planarity and closed outer k-quasi-planarity using monadic second-order logic (MSO$_2$; see Section 2.4) and Courcelles' Theorem (stated as Theorem 2.1 in Section 2.4). First, we design MSO$_2$ formulas expressing crossing patterns of closed k-planar and closed k-quasi-planar drawings. Thus, using Observation 6.2, we can test full outer k-planarity (full outer k-quasi-planarity) of a graph G by testing its biconnected components for closed outer k-planarity (closed outer k-quasi-planarity) using the MSO$_2$ formulas. Based on this together with the Courcelles' theorem and the fact that outer k-planar graphs have bounded treewidth (see Proposition 8.5 of [WT07]) we give a linear time algorithm for testing full outer k-planarity.

The challenge in expressing outer k-planarity or outer k-quasi-planarity in MSO$_2$ is that MSO$_2$ does not allow quantification over sets of pairs of vertices which involve non-edges. Namely, it is unclear how to express a set of pairs that forms the circular order of vertices on the boundary of our convex drawing. However, if this circular order forms a Hamiltonian cycle in our graph, i.e., the given graph is closed, then we can indeed express this in MSO$_2$. With the edge set of a Hamiltonian cycle of our graph in hand, we can then ask that this cycle was chosen in such a way that the other edges satisfy either k-planarity or k-quasi-planarity.

Any formula presented here assumes that a graph G is given and uses edges, vertices and incidences of G. In the following, e, f are edges, F is a set of edges, u, v are vertices and U a set of vertices (also including sub- and superscripted variants). In addition to the quantifiers above we also use a logical shorthand $\exists^{=x}$ for the existence of exactly x elements satisfying the property, that all are unequal and that no $x + 1$ such elements exist.

The following formula allows us to describe connectedness of subgraphs induced by an edge set F.

$$\text{Connected-Edges}(F) \equiv (\forall F' \subset F)[\exists e, f : e \in F' \wedge f \in F \setminus F'] \wedge \Big($$
$$(\exists f, e, e^* \in F : e \in F' \wedge f \notin F')(\exists u, v)$$
$$[I(u, f) \wedge I(v, e) \wedge I(v, e^*) \wedge I(u, e^*)]\Big)$$

It states that for every proper subset F' of our edge set F, we can find three edges e, f, e^* – one in F', one not in F' and one connecting the two.

These formulas are used to describe Hamiltonicity of G. Cycle-Set implies that the edges of F form cycles, Cycle implies maximality of the cycle and Span forces the cycle to have an edge incident to every vertex of G.

$$\text{Cycle-Set}(F) \equiv (\forall e)\Big[e \in F \Rightarrow (\exists^{=2} f)[f \in F \wedge e \neq f \wedge (\exists v)[I(e, v) \wedge I(f, v)]]\Big]$$
$$\text{Cycle}(F) \equiv \text{Cycle-Set}(F) \wedge \text{Connected-Edges}(F)$$
$$\text{Span}(F) \equiv (\forall v)(\exists e)[e \in F \wedge I(e, v)]$$
$$\text{Hamiltonian}(F) \equiv [\text{Cycle}(F) \wedge \text{Span}(F)]$$

Vertex-Partition implies the existence of a partition of the vertices of G into k disjoint subsets.

$$\text{Vertex-Partition}(U_1, \ldots, U_k) \equiv (\forall v)\left[\left(\bigvee_{i=1}^{k} v \in U_k\right) \wedge \left(\bigwedge_{i \neq j} \neg(v \in U_i \wedge v \in U_j)\right)\right]$$

For a closed outer k-planar or closed outer k-quasi-planar graph G, we want to express that two edges e and e_i cross. To this end, we assume that there is a Hamiltonian cycle E^* of G that defines the outer face. We partition the vertices of G into three subsets C, A, and B, as follows: C is the set containing the endpoints of e, whereas A and B are connected subgraphs on the remaining vertices that use only edges of E^*. In this way, we partition the vertices of G into two sets, one left and the other one right of e. For such a partition, e_i must cross e whenever e_i has one endpoint in A and one in B.

$$\text{Crossing}(E^*, e, e_i) \equiv (\forall A, B, C)\Big[\big(\text{Vertex-Partition}(A, B, C)$$
$$\wedge (I(e, x) \leftrightarrow x \in C) \wedge \text{Connected}(A, E^*) \wedge \text{Connected}(B, E^*)\big)$$
$$\rightarrow (\exists a \in A)(\exists b \in B)[I(e_i, a) \wedge I(e_i, b)]\Big]$$

Now we can describe the crossing patterns for closed outer k-planarity and closed outer k-quasi-planarity as follows:

$$\text{Closed Outer } k\text{-Planar}_G \equiv (\exists E^*)\Big[\text{Hamiltonian}(E^*) \wedge$$

$$(\forall e)\left[(\forall e_1,\ldots,e_{k+1})\left[\left(\bigwedge_{i=1}^{k+1} e_i \neq e \wedge \bigwedge_{i\neq j} e_i \neq e_j\right) \to \bigvee_{i=1}^{k+1} \neg\text{Crossing}(E^*,e,e_i)\right]\right]$$

Here we insist that G is Hamiltonian and that, for every edge e and any set of $k + 1$ distinct other edges, at least one among them does not cross e.

$$\text{Closed Outer } k\text{-Quasi-Planar}_G \equiv (\exists E^*)\left[\text{Hamiltonian}(E^*)\wedge\right.$$

$$(\forall e_1,\ldots,e_k)\left[\left(\bigwedge_{i\neq j} e_i \neq e_j\right) \to \bigvee_{i\neq j} \neg\text{Crossing}(E^*,e_i,e_j)\right]\right]$$

Again, we insist that G is Hamiltonian and further that, for any set of k distinct edges, there is at least one pair among them that does not cross.

The formulas above give us the following.

Theorem 6.5. *Closed outer k-planarity and closed outer k-quasi-planarity can be expressed in MSO_2 with a formula whose size depends only on k.*

Theorem 6.6. *We can test whether a graph G is full outer k-planar in linear time.*

Proof: Recall that in a full outer k-planar drawing there is no crossing on the outer boundary of the drawing and each biconnected component of the graph with such a drawing is a closed outer k-planar graph (Observation 6.2). Thus, in order to test full outer k-planarity for a given graph G it suffices to test whether each of its biconnected components admits a closed outer k-planar drawing. We can brake up G into biconnected components in linear time by obtaining the set of cutvertices of the graph in linear time. Checking each biconnected component for closed outer k-planarity can be done via the above MSO_2 formula in time linear in the size of the component. The formula also guarantees that the Hamiltonian cycle (if present) is placed on the outer boundary of the drawing of each component. Putting together the individual drawings of the components crossing free by reidentifying the cutvertices can also be done in linear time. Thus, the total runtime is linear in the size of the input graph G. □

Alternatively, we could encode the recognition of full outer k-planar and k-quasi planar graphs directly using an MSO_2 formula. This, however, will be more time consuming than the above approach.

6.4   SAT Formula for Testing Outer Quasi-Planarity of a Graph

In this section, we describe a logical formula for testing whether a given graph is outer quasi-planar. We present the formula in first-order logic. After transformation to Boolean logic, we solve the formula using MiniSat [ES18].

A quasi outer-planar embedding corresponds to a circular order of the vertices. If we cut a circular order at some vertex to turn the circular into a linear order, the edge crossing pattern remains the same. Therefore, we look for a linear order. For any pair of vertices $u \neq v \in V$, we introduce a Boolean variable $x_{u,v}$ that expresses that vertex u is before v in the linear order. In addition, for any pair of edges $e \neq e' \in E$ we introduce a Boolean variable $y_{e,e'}$ that expresses that edge e crosses edge e'. Now we list the clauses of our SAT formula.

$$x_{u,v} \wedge x_{v,w} \Rightarrow x_{u,w} \qquad \text{for each } u \neq v \neq w \in V; \tag{6.2}$$

$$x_{u,v} \Leftrightarrow \neg x_{v,u} \qquad \text{for each } u \neq v \in V; \tag{6.3}$$

$$x_{u,u'} \wedge x_{u',v} \wedge x_{v,v'} \Rightarrow y_{e,e'} \qquad \text{for each } e = (u,v) \neq e' = (u',v') \in E; \tag{6.4}$$

$$\neg(y_{e_1,e_2} \wedge y_{e_1,e_3} \wedge y_{e_2,e_3}) \qquad \text{for each } e_1, e_2, e_3 \in E \text{ with different endpoints.} \tag{6.5}$$

The first two sets of clauses describe the linear order. Clause (6.2) ensures transitivity, and clause (6.3) anti-symmetry. Clause (6.4) realizes the intended meaning of variable $y_{e,e'}$. Finally, clause (6.5) ensures that no three edges pairwise cross.

6.5   SAT Formula for Testing the Page Number of a Graph

A similar SAT solver has been implemented by Pupyrev [Pup17]. For a given graph G and integer $k > 0$, we provide a SAT formula that has a satisfying truth assignment if and only if G has page number k. We find a linear order of the vertices that corresponds to a k-page embedding. For every pair of vertices $u \neq v \in V$, we introduce a Boolean variable $x_{u,v}$ (as in Section 6.4) that expresses that u is before v in the linear order. For every edge $e \in E$ and page $i \in \mathcal{P} = \{1, \ldots k\}$, we introduce a Boolean variable $p_{i,e}$ that expresses that edge e is on page i. Now we list the clauses of our SAT formula.

$$x_{u,v} \wedge x_{v,w} \Rightarrow x_{u,w} \qquad \text{for each } u \neq v \neq w \in V;$$

$$x_{u,v} \Leftrightarrow \neg x_{v,u} \qquad \text{for each } u \neq v \in V;$$

$$\bigvee_{i \in \mathcal{P}} p_{i,e} \qquad \text{for each } e \in E; \tag{6.6}$$

$$\neg(p_{i,e} \wedge p_{j,e}) \qquad \text{for each } i \neq j \in \mathcal{P}, e \in E; \tag{6.7}$$

$$p_{i,e} \wedge_{i,e'} \Rightarrow \neg(x_{u,u'} \wedge x_{u',v} \wedge x_{v,v'}) \qquad \text{for each } i \in \mathcal{P}, e \neq e' \in E: \tag{6.8}$$
$$e = (u,v) \text{ and } e' = (u',v').$$

The first two sets of clauses are the same as Clauses (6.2)–(6.3) since they describe the linear order. Clauses (6.6)–(6.7) guarantee that every edge is on a unique page. Clause (6.8) ensures that two edges do not cross on the same page.

6.6   Discussions and Open Problems

Every planar graph is outer 5-quasi-planar because planar graphs have page number 4 [Yan89] (planar graphs that require 4 pages have also been discovered recently [Yan20, BKK+20]). There are also planar graphs that are not outer 3-quasi-planar. It is open whether every planar graph is outer 4-quasi-planar.

So far the trivial upper bound on degeneracy of outer k-quasi-planar graphs comes from the edge density, that is, every outer k-quasi-planar graph has at most $2(k-1)n - \binom{2k-1}{2}$

edges [CP92], thus, every outer k-quasi-planar graph has a vertex of degree at most $4(k-1)$. Can this bound be improved?

We now discuss the relation between our crossing-restricted convex drawings and the class of intersection graphs of chords of a circle, i.e., *circle graphs*. Such representations are called *chord diagrams*. Here, a convex drawing D of a graph G can be seen as a chord diagram and as such provides a corresponding graph H where each adjacency between two vertices corresponds to a crossing between the edges of our drawing. Independent sets in H correspond to collections of pairwise non-crossing edges in D, i.e., outerplanar sub-drawings of D. Thus, k-coloring H corresponds to partitioning D into edge sets $E_1, \dots, E_k$ such that each sub-drawing of D formed by the edges of E_i is outerplanar. That is, the partition $E_1, \dots, E_k$ forms a book embedding of G with k pages. So, k-coloring the chord diagram provides a k-page book embedding of G. Interestingly, it is NP-complete to test whether a chord diagram can be 4-colored [GJGP80], but testing whether it can be 3-colored [Wik16] is still open. On the other hand, circle graphs are χ-*bounded* [KK97], i.e., the chromatic number $\chi(G)$ of a circle graph G is bounded by a function of the *clique number* $\omega(G)$ of G, that is, the number of vertices in the maximum clique of G. The best known bound is $21 \cdot 2^\omega - 24\omega - 24$ and is due to Černý [Č07]. In particular, this means that every outer k-quasi planar drawing can be partitioned into $21 \cdot 2^{k-1} - 24k$ pages (since we cannot have k mutually crossing edges, i.e., there is no k-clique in the corresponding intersection graph). For quasi-planar graphs ($k = 3$) however a better bound is known. Ageev [Age96] showed that any triangle-free circle graph has chromatic number at most 5. Because for a fixed drawing of an outer quasi-planar graph its corresponding circle graph is triangle-free it has chromatic number at most 5, and thus, we can embed the outer quasi-planar graph in a book with 5 pages. An immediate open question is to improve this bound on the page number.

Ageev [Age96] constructed a circle graph G_{Ageev} with $\chi(G_{\text{Ageev}}) = 5$. The drawing of the outer quasi-planar graph G corresponding to the circle graph G_{Ageev} cannot be embedded on four pages because the circle graph has chromatic number 5. It turns out, however, that there exists a linear order of the vertices under which G can be embedded on four pages, even if we add edges to make it maximal, but there does not exist such an order that the drawing in addition is outer quasi-planar. We have verified this experimentally by constructing a logical formula that tests outer quasi-planarity and 4-page embeddability at the same time; see Sections 6.4 and 6.5.

7

Bundled Crossing Minimization in Circular Layouts

So how can we tell that a drawing with crossings is close to planar, in other words, beyond-planar? In Chapter 5 we identified such drawings as drawings where edges have small complexity and cross at right angles only and in Chapter 6 we considered a drawing beyond-planar if it does not have too many crossings, that is, at most k crossings per edge or at most k pairwise crossing edges. But for large and dense graphs traditional node–link diagrams tend to contain a lot of crossings [ACNS82]. For this reason, Holten [Hol06] introduced *bundled drawings*, where edges that are close together and roughly go into the same direction are drawn using Bézier curves such that the grouping becomes visible. Due to the practical effectiveness of this approach, it has quickly been adopted by the InfoVis community [CZQ+08, PNBH16, GHNS11, HET12, HEF+14]. However, recently bundled drawings have also attracted study from a theoretical point of view [AFP16, FHSV16, FPW15, vDFF+17].

The natural family of beyond-planar graphs in the bundled variant is graphs with few crossings of bundles or bundled crossings; see Definition 7.1 for the formalization of a bundled crossing. In fact, in his survey on crossing minimization, Schaefer lists the bundled crossing minimization problem as a variant of the crossing minimization problem and suggests to study it [Sch17, page 35].

Related work.   Fink et al. [FPW15] considered bundled crossings (which they called block crossings) in the context of drawing metro maps. A metro network is a planar graph where vertices are stations and metro lines are simple paths in this graph. These paths representing metro lines can share edges. They enter an edge at one endpoint in some linear order, follow the edge as x-monotone curves (considering the edge as horizontal), and then leave the edge at the other endpoint in some linear order. In order to improve the readability of metro maps, the authors suggested to bundle crossings. The authors then studied the problem of minimizing bundled crossings in such metro maps. Fink et al. also introduced *monotone* bundled crossing minimization where each pair of lines can intersect at most once. Later, Fink et al. [vDFF+17] applied the concept of bundled crossings to drawing storyline visualizations. A storyline visualization is a set of x-monotone curves where the x-axis represents time in a story. Given a set of *meetings* (subsets of the curves that must be consecutive at given points in time), the task is to find a drawing that realizes the meetings and minimizes the number of bundled crossings. Fink et al. showed that, in this setting, minimizing bundled crossings is fixed-parameter tractable (FPT) in the number of curves and can be approximated in a restricted case. Van Dijk et al. [vDLMW18] gave ILP and SAT formulations of the problem and evaluated these experimentally.

Our research builds on recent works of Fink et al. [FHSV16] and Alam et al. [AFP16], who extended the notion of bundled crossings from sets of x-monotone curves to general drawings of graphs. We discuss their results in more detail soon.

The *degenerate crossing number* is defined by allowing more than two edges to intersect at the same point; several variants (one of which is also called the genus crossing number) have been studied [Moh09, PT09, AP13, SŠ15]. The degenerate crossing number and the bundled crossing number might look completely different, but it turns out that the degenerate crossing number is closely related to the non-orientable genus [Moh09] while the bundled crossing number is closely related to the orientable genus as we will see.

Notation and definitions.   In graph drawing, it is common to define a drawing of a graph as a function that maps vertices to distinct points in the plane and edges to Jordan arcs that connect the corresponding points. In this chapter, we are less restrictive; we sometimes allow edges to self-intersect. However, we forbid any three edges to share the same point. We will often identify vertices with their points and edges with their curves. Moreover, we assume that each pair of edges shares at most a finite number of points, that edges can touch (that is, be tangent to) each other only at endpoints, and that any point of the plane that is not a vertex is contained in at most two edges. A drawing is *simple* if any two edges intersect at most once and no edge self-intersects. We consider both simple and non-simple drawings; look ahead at Figure 7.2 for a simple and a non-simple drawing of $K_{3,3}$.

Definition 7.1 (Bundled Crossing). Let D be a drawing, not necessarily simple, and let $I(D)$ be the set of intersection points among the edges (not including the vertices) in D. We say that a *bundling* of D is a partition of $I(D)$ into *bundled crossings*, where a set $B \subseteq I(D)$ is a bundled crossing if the following holds (see Figure 7.1).

- B is contained in a closed region $R(B)$ of the plane whose boundary consists of four Jordan arcs $\tilde{e}_1$, $\tilde{e}_2$, $\tilde{e}_3$, and $\tilde{e}_4$ that are pieces of edges e_1, e_2, e_3, and e_4 in D (with $\tilde{e}_i = e_i \cap R(B)$ for $i \in \{1, 2, 3, 4\}$).

- The pieces of the edges cut out by the region $R(B)$ can be partitioned into two sets $\tilde{E}_1$ and $\tilde{E}_2$ such that $\tilde{e}_1, \tilde{e}_3 \in \tilde{E}_1$, $\tilde{e}_2, \tilde{e}_4 \in \tilde{E}_2$, and each pair of edge pieces in $\tilde{E}_1 \times \tilde{E}_2$ has exactly one intersection point in $R(B)$, whereas no two edge pieces in $\tilde{E}_1$ intersect and no two edge pieces in $\tilde{E}_2$ intersect.

Our definition is similar to that of Alam et al. [AFP16] but defines the Jordan region $R(B)$ more precisely. We call the sets $\tilde{E}_1$ and $\tilde{E}_2$ of edge pieces *bundles* and the Jordan arcs $\tilde{e}_1, \tilde{e}_3 \in \tilde{E}_1$ and $\tilde{e}_2, \tilde{e}_4 \in \tilde{E}_2$ *frame arcs* of the bundles $\tilde{E}_1$ and $\tilde{E}_2$, respectively. For simple drawings, we accordingly call the edges that bound the two bundles of a bundled crossing *frame edges*. We say that a bundled crossing is *degenerate* if at least one of the bundles consists of only one edge piece; see Figure 7.1b. In this case, the region of the plane associated with the crossing coincides with that edge piece. In particular, any point in $I(D)$ by itself is a degenerate bundled crossing. Hence, every drawing admits a trivial bundling.

We use bc(G) to denote the *bundled crossing number* of a graph G, i.e., the smallest number of bundled crossings over all bundlings of all simple drawings of G. When we do not insist on simple drawings, we denote the corresponding number by bc$'(G)$. In the circular setting, where vertices are required to lie on the boundary of a disk and edges inside this disk, we consider the analogous *circular bundled crossing numbers* bc$^\circ(G)$ and bc$^{\circ\prime}(G)$ of a graph G. If, in addition, the vertices are required to be in a prescribed circular order π, we consider

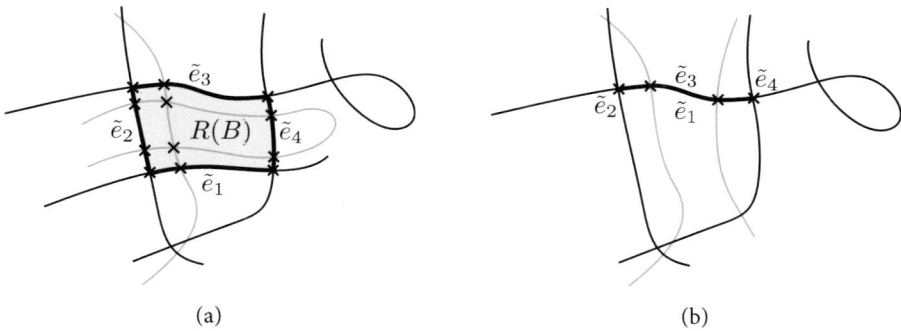

Figure 7.1: (a) A non-degenerate bundled crossing B and (b) a degenerate bundled crossing B'; crossings belonging to a bundled crossing are marked with crosses.

the circular bundled crossing number with a *fixed vertex order* π and denote this number as $\mathrm{bc}^{\circ}(G, \pi)$.

By $\mathrm{bc}(G, D)$ we denote the bundled crossing number of a specific simple drawing D of G. We say *fixed* drawing for this case. Similarly, by $\mathrm{bc}'(G, D)$ we denote the bundled crossing number of a specific, not necessarily simple drawing D of a graph G. By $\mathrm{bc}^{\circ}(G, D)$ we denote the bundled crossing number of a simple circular drawing D of a graph G. As Fink et al. [FHSV16] observed in this variant of the problem, we can assume the graph to be a matching. They [FHSV16] used "embedding" and $\mathrm{bc}(\mathcal{E})$ where we use "drawing" and $\mathrm{bc}(G, D)$, respectively.

Fink et al. [FHSV16] showed that it is NP-hard to compute the minimum number $\mathrm{bc}(G, D)$ of bundled crossings that a given drawing D of a graph G can be partitioned into. They also showed that this problem generalizes the problem of partitioning a rectilinear polygon with holes into the minimum number of rectangles, and they exploited this connection to construct a 10-approximation for computing the number $\mathrm{bc}^{\circ}(G, D)$ of bundled crossings in the case of a circular drawing. They left open the computational complexity of the general and the circular bundled crossing number for the case that the drawing is not fixed.

Alam et al. [AFP16] showed that $\mathrm{bc}'(G)$ equals the orientable genus of G, which in general is NP-hard to compute [Tho89]. They also showed that there is a graph G with $\mathrm{bc}'(G) \neq \mathrm{bc}(G)$ by proving that $\mathrm{bc}'(K_6) = 1 < \mathrm{bc}(K_6)$. As it turns out, the two problem variants differ in the circular setting, too (see Figure 7.2 and Observation 7.2). For computing $\mathrm{bc}(G)$ and $\mathrm{bc}^{\circ}(G)$, Alam et al. [AFP16] gave an algorithm whose approximation factor depends on the density of the graph. They posed the existence of an FPT algorithm for $\mathrm{bc}^{\circ}(G)$ as an open question.

Our contribution.   As some graphs G have $\mathrm{bc}'(G) \neq \mathrm{bc}(G)$ (see Figure 7.2), Fink et al. [FHSV16] posed the complexity of computing the bundled crossing number $\mathrm{bc}(G)$ of a given graph G as an open problem. We settle this in Section 7.1 as follows:

Theorem 7.1.   *Given a graph G, it is NP-hard to compute $\mathrm{bc}(G)$.*

Our main result, which we prove in Section 7.2, resolves an open question of Alam et al. [AFP16] concerning the fixed-parameter tractability of bundled crossing minimization in circular layouts as follows:

Theorem 7.2. *There is a computable function f such that, for any n-vertex graph G and integer k, we can check, in $O(f(k)n)$ time, whether $\mathrm{bc}^\circ(G) \le k$, i.e., whether G admits a circular layout with k bundled crossings. Within the same time bound, such a layout can be computed.*

To prove this, we again, as in Chapter 6, use an approach similar to that of Bannister and Eppstein [BE18] for 1-page crossing minimization (that is, edge crossing minimization in a circular layout). Bannister and Eppstein observe that the set of crossing edges of a circular layout with k edge crossings of a graph G forms an arrangement of curves that partition the drawing into $O(k)$ subgraphs, each of which occurs in a distinct face of this arrangement. The subgraphs are obviously outerplanar. This means that G has bounded treewidth. So, by enumerating all ways to draw the crossing edges of a circular layout with k edge crossings, and, for each such way, expressing the edge partition problem (into crossing edges and outerplanar components) in extended monadic second-order logic (MSO$_2$; see Section 2.4), Courcelle's Theorem [Cou90] (stated as Theorem 2.1 in Section 2.4) can be applied (leading to fixed-parameter tractability).

The difficulty in using this approach for bundled crossing minimization is in showing how to partition the graph into a set of $O(k)$ "crossing edges" (our analogy will be the frame edges) and a collection of $O(k)$ outerplanar graphs. This is where we exploit the connection to genus. Moreover, constructing an MSO$_2$ formula is somewhat more difficult in our case due to the more complex way our regions interact with our special set of edges.

Again using the above-mentioned connection, here between genus and the circular bundled crossing number $\mathrm{bc}^{\circ\prime}$, we can decide whether $\mathrm{bc}^{\circ\prime}(G) = k$ in $2^{O(k)}n$ time. In other words, if non-simple drawings are allowed, the problem is also FPT in k; see Section 7.2 (Theorem 7.4).

We also consider the setting where we are given a drawing and the task is to bundle the existing edge crossings into as few bundled crossings as possible, that is, computing $\mathrm{bc}(G, D)$ for a given drawing D of a graph G. We show in Section 7.3 that we can use an algorithm of Marx and Philipczuk [MP15, Theorem 1.3] (see page 102) to test whether $\mathrm{bc}(G, D) \le k$ in $m^{O(\sqrt{k})}$ time for any simple drawing D with m edges. This yields an FPT-algorithm for testing whether $\mathrm{bc}^\circ(G, D) \le k$ in $2^{O(\sqrt{k}\log k)} + O(m)$ time and for testing whether $\mathrm{bc}^\circ(G, \pi) \le k$ in $2^{O(k^2)} + O(m)$ time, improving on an $\left(2^{O(k^2 \log k)} + O(m)\right)$-time algorithm of Alam et al. [AFP16].

In Section 7.4 we consider storyline visualizations. In contrast to the above results, the storyline literature considers the number of characters m to be small and the number of crossings to be large. (Recall that storyline visualizations are non-simple.) We show that computing the bundled crossing number $\mathrm{bc}^s(D)$ of a given storyline visualization D can be done in $O(\varphi^{2m}\mathrm{poly}(m+c))$ time, where c is the number of crossings in D and φ is the golden ratio. Note that this is fixed parameter tractable in m.

For an overview of existing and new results see Table 7.1.

Table 7.1: Algorithmic and complexity results concerning bundled crossing minimization for an m-edge graph G with restrictions such as vertex order π, drawing D, edge density δ, and k bundled crossings. We omit polynomial terms, and φ is the golden ratio. Our results are in boldface.

General layout			Circular layout		
bc(G)	$\frac{6\delta}{\delta-3}$-approx. for $\delta>3$ [AFP16]		bc$^\circ$(G)	$\frac{6\delta}{\delta-2}$-approx. for $\delta>2$	[AFP16]
	NP-hard	(Thm. 7.1)		**FPT**	(Thm. 7.2)
bc$'$(G)	NP-hard	[FHSV16]	bc$^{\circ\prime}$(G)	**FPT**: $2^{O(k)}$	(Thm. 7.4)
bc(G,D)	NP-hard	[FHSV16]	bc$^\circ$(G,D)	10-approximation	[FHSV16]
	XP: $m^{O(\sqrt{k})}$	(Thm. 7.5(b))		**FPT**: $k^{O(\sqrt{k})}$	(Thm. 7.5(c))
bc$'$(G,D)	**XP**: $c^{O(\sqrt{k})}$	(Thm. 7.5(a))			
Storyline layout			bc$^\circ$(G,π)	16-approx., **FPT**: $k^{O(k^2)}$	[AFP16]
bc^s(D)	**FPT**: φ^{2m}	(Thm. 7.7)		**FPT**: $2^{O(k^2)}$	(Thm. 7.5(d))

7.1 Computing bc(G) Is NP-Hard

For a given graph G, finding a drawing with the fewest bundled crossings resembles computing the *orientable genus* g(G) of G, that is, computing the fewest *handles* to attach to the sphere so that G can be drawn on the resulting surface without any crossings. In fact, Alam et al. [AFP16] showed that bc$'$(G) = g(G). Thus, deciding whether bc$'$(G) = k for some k is NP-hard and FPT in k since the same holds for deciding whether g(G) = k [Tho89, Moh99, KMR08].

Theorem 7.3 ([AFP16]). *For every graph G with genus k, it holds that* bc$'$(G) = k.

To show this, Alam et al. [AFP16] first showed that a drawing with k bundled crossings can be lifted onto a surface of genus k, and thus bc$'$(G) $\geq$ g(G):

Observation 7.1 ([AFP16]). *A drawing D with k bundled crossings can be* lifted *onto a surface of genus k via a one-to-one correspondence between bundled crossings and handles, i.e., at each bundled crossing, we attach a handle for one of the two edge bundles, thus providing a crossing-free lifted drawing; see Figure 7.8.*

Then, to see that bc$'$(G) $\leq$ g(G), Alam et al. [AFP16] used the *fundamental polygon* representation (or *polygonal schema*) [de 17] of a drawing on a genus-g surface. More precisely, the sides of the polygon are numbered in circular order $a_1, b_1, a_1', b_1', \ldots, a_g, b_g, a_g', b_g'$; for $1 \leq k \leq g$, the pairs (a_k, a_k') and (b_k, b_k') of sides are identified in opposite direction, meaning that an edge leaving side a_k appears on the corresponding position of side a_k'; see Figure 7.3 and Figure 7.4a for an example showing K_6 drawn in a fundamental square, which models a drawing on the torus. In such a representation, all vertices lie in the interior of the fundamental polygon and all edges leave the polygon avoiding vertices of the polygon. Alam et al. [AFP16] showed that such a representation can be transformed into a non-simple bundled drawing with g many bundled crossings. It is not clear, however, when such a representation can be transformed into a simple bundled drawing with g bundled crossings, as this transformation can produce drawings with self-intersecting edges and pairs of edges crossing multiple times, e.g., Alam et al. [AFP16, Lemma 1] showed that $bc(K_6) = 2$ while $bc'(K_6) = g(K_6) = 1$.

We now show that computing the bundled crossing number remains NP-hard for simple drawings.

Proof of Theorem 7.1: Let G' be the graph obtained from G by subdividing each edge $O(|E(G)|^2)$ times. We reduce from the NP-hardness of computing the genus $g(G)$ of G by showing that $bc(G') = g(G)$, with Observation 7.1 in mind.

Consider the embedding of G onto the genus-$g(G)$ surface. By a result of Lazarus et al. [LPVV01, Theorem 1], we can construct a fundamental polygon representation of the embedding so that its boundary intersects with edges of the graph $O(g(G)|E(G)|)$ times. Note that each edge piece outside the polygon between the sides of the polygon (a_k, a'_k) intersects each other edge piece outside the polygon between the sides of the polygon (b_k, b'_k) at most once and does not have any other intersection points; see Figure 7.3. We then subdivide the edges by adding a vertex to each intersection of an edge with the boundary of the fundamental polygon. This process of subdividing edges ensures that no edge intersects itself or intersects another edge more than once in the corresponding drawing of the graph on the plane; hence, the drawing is simple. Since $g(G) \leq |E(G)|$, by subdividing edges further whenever necessary, we obtain a drawing of G'. Our subdivisions keep the integrity of all bundled crossings, so $bc(G') \leq g(G)$. On the other hand, since subdividing edges does not affect the genus, $g(G) = g(G') = bc'(G') \leq bc(G')$. □

7.2 Computing $bc^{\circ'}(G)$ and $bc^{\circ}(G)$ Is FPT

We now consider circular layouts, where vertices are placed on a circle and edges are routed inside the circle. We note that $bc^{\circ}(G)$ and $bc^{\circ'}(G)$ can be different.

Observation 7.2. $bc^{\circ'}(K_{3,3}) = 1$ *but* $bc^{\circ}(K_{3,3}) > 1$.

Proof: Let $V(K_{3,3}) = \{a, b, c\} \cup \{a', b', c'\}$. A drawing with $bc^{\circ'}(K_{3,3}) = 1$ is obtained by placing the vertices a, a', b, b', c, c' in clockwise order around a circle; see Figure 7.2b. If a graph G has $bc^{\circ}(G) = 1$ then G is planar because we can embed edges for one bundle outside the circle. Hence, $bc^{\circ}(K_{3,3}) > 1$. □

Similarly to computing $bc'(G)$, we compute $bc^{\circ'}(G)$ via computing genus. To show this we first prove the following.

Lemma 7.1. *Given a graph G, let $G^\star$ be the graph obtained from G by adding a new vertex $v^\star$ adjacent to every vertex of G. Then $bc^{\circ'}(G) = g(G^\star)$.*

Proof: Let G be the given graph with $V = V(G)$ and $E = E(G)$. Similarly as in [AFP16, Theorem 1], it is easy to see that $bc^{\circ'}(G)$ is an upper bound for the genus of $G^\star$, because, according to Observation 7.1, we can lift any circular drawing of G onto a surface S of genus $bc^{\circ'}(G)$ and then we can add $v^\star$ using the outside of the circle. Clearly, this produces a crossing-free drawing of $G^\star$ on the surface S.

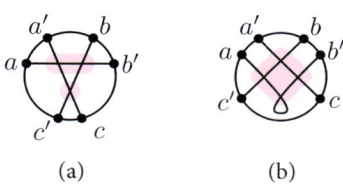

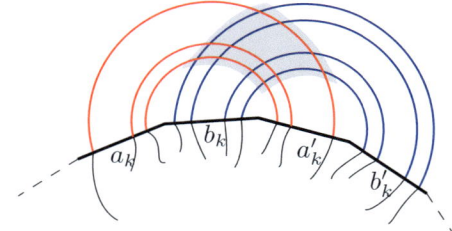

Figure 7.2: $bc°(K_{3,3}) \neq bc°′(K_{3,3})$; see Observation 7.2

Figure 7.3: A single bundled crossing outside the fundamental polygon [AFP16, Figure 3].

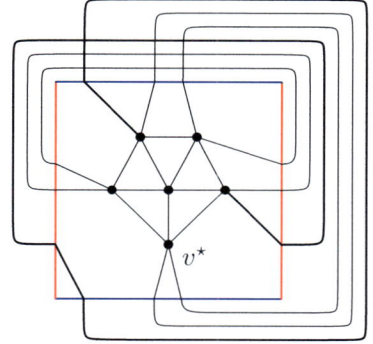

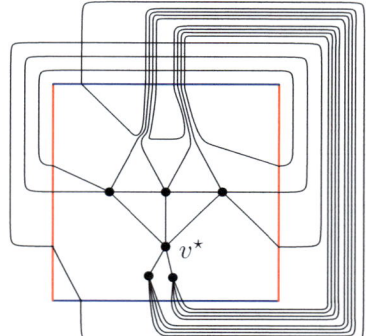

(a) K_6 drawn in a fundamental square; the self-intersecting edge is bold [AFP16, Figure 2]

(b) modifying the representation

Figure 7.4: Obtaining a circular drawing with k bundled crossings of G from the embedding of $G^\star$ on a surface of genus k.

It remains to show that given a crossing-free drawing of $G^\star$ on a surface of genus k, we can construct a circular drawing of G with at most k bundled crossings. Consider a drawing of $G^\star$ on a surface $\mathcal{S}$ of genus k.

We use the fundamental polygon representation [AFP16, Theorem 1] to the drawing of $G^\star$ on the surface $\mathcal{S}$ of genus k; see Figure 7.4a. Then we modify this representation so that all the neighbors $N(v^\star)$ of $v^\star$ in $G^\star$ are placed in an ϵ-neighborhood of $v^\star$. We now explain the modification in more detail. Consider all the edges incident to $v^\star$ in the representation and drag each neighbor u of $v^\star$ along the edge $uv^\star$ (as illustrated in Figure 7.4b) until it reaches the ϵ-neighborhood $N(v^\star)$ of $v^\star$. Since for each $u \in N(v^\star)$ the edges $uw \in E$ with $w \neq v^\star$ are bundled together at the position where u was in the representation and dragged together with u along the edge $uv^\star$, this does not change the number of bundled crossings. Since all the vertices are located on the boundary of the ϵ-neighborhood of $v^\star$ in the modified representation, all the edges between $v^\star$ and $V \setminus v^\star$ are drawn inside the polygon. After removing the vertex $v^\star$ from the representation, we obtain a circular drawing of G with at most k bundled crossings. □

Theorem 7.4. *Testing whether* $\mathrm{bc}^{\circ'}(G) = k$ *can be done in* $2^{k^{O(1)}} n$ *time.*

Proof: By Lemma 7.1, $\mathrm{bc}^{\circ'}(G) = g(G^\star)$, where $G^\star$ is a graph with a vertex $v^\star$ adjacent to every vertex of G. Applying the $\left(2^{g^{O(1)}} n\right)$-time algorithm for computing genus [KMR08] completes the proof. □

To prove our main result (Theorem 7.2) we develop an algorithm that tests whether $\mathrm{bc}^{\circ}(G) = k$ in FPT time with respect to k. As in Chapter 6 our algorithm is inspired by the recent work on circular layouts with at most k crossings [BE18]. In Chapter 6 as well as in the algorithm of Bannister and Eppstein [BE18], it is first observed that the graphs admitting such circular layouts have treewidth $O(k)$, and then algorithms are developed using Courcelle's theorem (see Theorem 2.1 in Section 2.4), which establishes that expressions in MSO_2 logic can be evaluated efficiently. (For the definition of treewidth see Section 2.1 and for formalization of MSO_2 logic see Section 2.4.)

We proceed along the lines of Bannister and Eppstein [BE18], who used a similar approach to show that edge crossing minimization in a circular layout is in FPT (as mentioned in the introduction). We start by very carefully describing a surface (in the spirit of Observation 7.1) onto which we will lift our drawing. We will then examine the structure of this surface (and our algorithm) for the case of one bundled crossing and finally for k bundled crossings.

7.2.1 Constructing the Surface Determined by a Bundled Drawing

Consider a bundled circular drawing D. Note that adding parallel edges to the drawing (i.e., making our graph a multi-graph) allows us to assume that every bundled crossing has four distinct frame edges and can be done without modifying the number of bundled crossings; see Figure 7.8. Each bundled crossing B defines a Jordan curve made up of the four Jordan arcs $\tilde{e}_1, \tilde{e}_2, \tilde{e}_3, \tilde{e}_4$ in clockwise order taken from its four frame edges $e_1, \ldots, e_4$, respectively, where (e_1, e_3) and (e_2, e_4) frame the two bundles and $e_i = v_{2i-1}v_{2i}$. Similarly to Observation 7.1, we can construct a surface $\mathcal{S}$ by creating a flat handle (note that this differs from the usual definition of a handle since our flat handles have a boundary) on top of D which connects $\tilde{e}_2$ to $\tilde{e}_4$ and doing so for each bundled crossing. We then lift the drawing D onto $\mathcal{S}$ by rerouting the edges of one of the bundles over its corresponding handle for each bundled crossing B obtaining the lifted drawing $D_\mathcal{S}$. To avoid the crossings in $D_\mathcal{S}$ of the frame edges that can occur at the foot of the handle of B, we can make the handle a bit wider and add *corner-cuts* (as illustrated in Figure 7.5b) to preserve the topology of the surface. Thus, $D_\mathcal{S}$ is crossing-free.

We now cut $\mathcal{S}$ into *components* (maximal connected subsets) along the frame edges and corner-cuts of each bundled crossing, resulting in a subdivision Ω of $\mathcal{S}$.

We use D_Ω to denote the sub-drawing of $D_\mathcal{S}$ on Ω, i.e., D_Ω is missing the frame edges since these have been cut out. We now consider the components of Ω. Notice that every edge of D_Ω is contained in one component of Ω. In order for a component s of Ω to contain an edge e of D_Ω, s must have both endpoints of e on its boundary. With this in mind we focus on certain components of Ω. Namely, we call a component a *region* if it contains a vertex of G on its boundary. Observe that a crossing in D which does not involve a frame edge corresponds, in D_Ω, to a pair of edges where one goes over a handle and the other goes underneath.

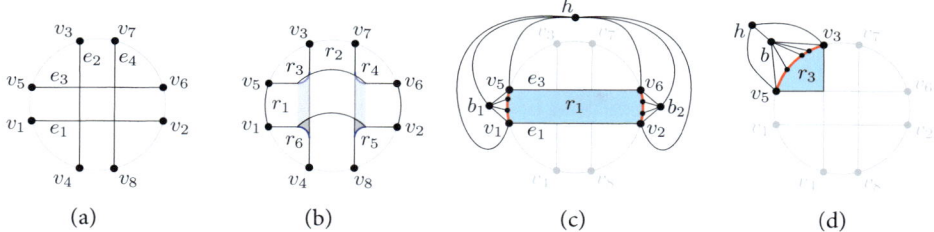

Figure 7.5: (a) Bundled crossing; (b) regions, corner-cuts in blue; (c),(d) the augmented graphs $G^*_{r_1}$ and $G^*_{r_3}$ consist of the edges of G_{r_1} and G_{r_3} (in the blue regions) as well as augmentation vertices and edges (drawn in black).

7.2.2 Recognizing a Graph with One Bundled Crossing

We now discuss how to recognize if an n-vertex graph G with $V = V(G)$ and $E = E(G)$ can be drawn in a circular layout with one bundled crossing. Consider a bundled circular drawing D of G consisting of one bundled crossing. The bundled crossing consists of two bundles, which are bounded by the set $F = \{e_1, e_2, e_3, e_4\}$ of frame edges. By $V(F)$ we denote the set of vertices incident to frame edges. Via the construction above, we obtain the subdivided surface Ω; see Figure 7.5. Let r_1 and r_2 be the regions that are each bounded by a pair of frame edges corresponding to one of the bundles, and let $r_3, \ldots, r_6$ be the regions each bounded by one edge from one pair and one from the other pair; see Figure 7.5b. These are all the regions of Ω. Since, as mentioned before, each of the non-frame edges of G (i.e., each $e \in E(G) \smallsetminus F$) along with its two endpoints is contained in exactly one of these regions, each component of $G \smallsetminus V(F)$ and each edge connecting it to vertices of $V(F)$ is drawn in D_Ω in some region of Ω. In this sense, for each region r of Ω, we use G_r to denote the subgraph of G induced by the components of $G \smallsetminus V(F)$ contained in r and the edges connecting them to vertices in $V(F)$. Additionally, each vertex of G is either incident to an edge in F (in which case it is on the boundary of at least two regions) or it is on the boundary of exactly one region.

Note that there are two types of regions: those in $\{r_1, r_2\}$ and those in $\{r_3, r_4, r_5, r_6\}$. Consider a region of the first type, say r_1; see Figure 7.5b. Observe that G_{r_1} is outerplanar. Moreover, G_{r_1} has a special outerplanar drawing where, on the boundary of r_1, we see (in clockwise order) the frame edge e_1, the vertices mapped to the (v_1, v_5)-arc, the frame edge e_3, and then the vertices mapped to the (v_6, v_2)-arc. We now describe how to augment G_{r_1} to a planar graph $G^*_{r_1}$ where in every planar embedding of $G^*_{r_1}$ the sub-embedding of G_{r_1} has this special outerplanar form; this augmentation may sound overly complicated, but is written as to easily generalize to more bundled crossings. The vertex set of $G^*_{r_1}$ is $V(G_{r_1}) \cup \{h, b_1, b_2\}$ where we call h *hub* vertex and b_1 and b_2 *boundary* vertices (one for each arc of the boundary of r_1 to which vertices can be mapped); see Figure 7.5c. The graph $G^*_{r_1}$ has four types of edges; the edges in $E(G_{r_1})$, edges that make h the hub of a wheel whose cycle is $C = (v_6, b_2, v_2, v_1, b_1, v_5, v_6)$, edges from b_1 to the vertices on the (v_1, v_5)-arc, and edges from b_2 to the vertices on the (v_6, v_2)-arc (both including the endpoints). Clearly, we can obtain a planar embedding of $G^*_{r_1}$ by drawing the elements of $G^*_{r_1} \smallsetminus G_{r_1}$ "outside" of the outerplanar drawing of G_{r_1} described before. Moreover, every planar embedding of $G^*_{r_1}$ contains an outerplanar embedding of G_{r_1}

that can be drawn in the special form needed to "fit" into r_1, in the sense that all of G_{r_1} lies (or can be put) inside the simple cycle C. (For example, if, say, b_1 is a cut vertex, the component hanging off b_1 can be embedded in the face (h, b_1, v_1, h). But then it can easily be moved into C. Similarly, a component that is incident only to v_5 and v_6 can end up in the face (h, v_5, v_6, h), but again, the component can be moved inside C.)

Similarly, for a region of the second type, say r_3, the graph G_{r_3} is outerplanar with a special drawing where all the vertices must be on the (v_3, v_5)-arc of the disk subtended by the two frame edges e_3 and e_2 bounding the region r_3. We augment r_3 similarly as r_1; see Figure 7.5d. For the augmented graph $G_{r_3}^*$, we add to G_{r_3} a boundary vertex b neighboring all vertices on the (v_3, v_5)-arc and a hub vertex h adjacent to v_3, b, and v_5. Again, $G_{r_3}^*$ is planar since G_{r_3} is outerplanar. Moreover, as b is adjacent to all vertices of G_{r_3}, in every planar embedding of $G_{r_3}^*$, G_{r_3} is embedded outerplanarly and, since b occurs on one side of the triangle $v_3 v_5 h$, the edge $v_3 v_5$ occurs on the boundary of this outerplanar embedding of G_{r_3}. Thus, each planar embedding of $G_{r_3}^*$ provides an outerplanar embedding of G_{r_3} that fits into r_3.

Note that each G_{r_i} fits into r_i because its augmented graph $G_{r_i}^*$ is planar ($\star$). Moreover, as outerplanar graphs have treewidth at most two [Mit79], each graph G_r is outerplanar, and adding the (up to) eight frame vertices raises the treewidth by at most 8, we see that the treewidth of G is at most 10. Namely, in order for G to have $bc^\circ(G) = 1$, it must have treewidth at most 10 (and this can be checked in linear time using an algorithm of Bodlaender [Bod96]).

To sum up, G has a circular drawing D with at most one bundled crossing because it has treewidth at most 10 and there exist (i) $\beta \leq 4$ frame edges $e_1, e_2, \ldots, e_\beta$ (this set is denoted F) and $v_1, \ldots, v_\xi$ frame vertices (this set is denoted V_F), (ii) a particular circular drawing D_F of frame edges, (iii) the drawing of the one bundled crossing B, and (iv) $\gamma \leq 6$ corresponding regions $r_1, \ldots, r_\gamma$ of the subdivided surface Ω so that the following properties hold. (Note that the frame vertices partition the boundary of the disk underlying Ω into $\eta \leq 8$ (possibly degenerate) arcs $p_1, \ldots, p_\eta$ where each such p_j is contained in a unique region r_{i_j} of Ω. Let $V_0(r_i)$ be the frame vertices incident to region r_i.)

1. $E(G)$ is partitioned into $E_0, E_1, \ldots, E_\gamma$, where $E_0 = \{f_1, \ldots, f_\beta\}$.

2. $V(G)$ is partitioned into $V_0, V_1, \ldots, V_\eta$, where $V_0 = \{w_1, \ldots, w_\xi\}$.

3. The mapping $w_i \leftrightarrow v_i$ and $f_i \leftrightarrow e_i$ defines an isomorphism between the subgraph of G formed by (V_0, E_0) and the graph (V_F, F).

4. For each $v \in V_0$ and each edge e incident to v, exactly one of the following conditions holds: (i) $e \in E_0$, or (ii) $e \in E_i$ and v is on the boundary of r_i.

5. For each $v \in V_j$, $j \neq 0$, all edges incident to v belong to E_{i_j}.

6. For each region r_i, let G_{r_i} be the graph $(V_0(r_i) \cup \bigcup_{j:i_j=i} V_j, E_i)$ (i.e., the subgraph that is to be drawn in r_i), and let $G_{r_i}^*$ be the corresponding augmented graph (i.e., as in $\star$ above). Each $G_{r_i}^*$ is planar.

We now describe the algorithm that tests whether a given graph G admits a simple circular drawing with one bundled crossing. First we check that the treewidth of G is at most 10. We then enumerate drawings of up to four edges in the circle. For the drawing D_F that is valid for the set F of frame edges of one bundled crossing, we define our surface and its regions (which makes the augmentation well-defined). We have intentionally phrased these properties so that it is clear that they are expressible in MSO_2 (see Section 7.2.4). The only property that is not obviously expressible is the planarity of $G_{r_i}^*$. To this end, recall that planarity is characterized by two forbidden minors (i.e., K_5 and $K_{3,3}$) and that, for every fixed graph H, there is an MSO formula MINOR_H so that for all graphs G, it holds that $G \vDash \text{MINOR}_H$ if and only if G contains H as a minor [CE12, Corollary 1.14]. Additionally, each $G_{r_i}^*$ can be expressed as an MSO-transduction (see Section 2.4) of G and our variables. Thus, by [CE12, Theorem 7.10] using the transduction and the MSO formula testing planarity, we can construct an MSO_2 formula ι so that when $G \vDash \iota$, $G_{r_i}^*$ is planar for every i. Therefore, Properties 1–6 can be expressed as an MSO_2 formula ψ and, by Courcelle's theorem, there is a computable function f such that we can test (in $O(f(\psi, t)n)$ time) whether $G \vDash \psi$ for an input graph G of treewidth at most t. Thus, since our graph has treewidth at most 10, applying Courcelle's theorem completes our algorihtm.

7.2.3 Recognizing a Graph with k Bundled Crossings

We now generalize the above approach to k bundled crossings. In a drawing D of G together with a solution consisting of k bundled crossings, there are $2k$ bundles making (up to) $4k$ frame edges F. As described above, these bundled crossings provide a surface S, its subdivision Ω, and the corresponding set of regions. The key ingredient above was that every region r contained an outerplane graph G_r. However, that is now non-trivial as our regions can go over and under many handles. To show this property, we first consider the following two partial drawings $D_A(p)$ and $D_B(p)$ of a matching with $p + 1$ edges $f_0, f_1, \ldots, f_p$ (see Figure 7.6) such that

- edge f_i crosses only $f_{i-1 \bmod p+1}$ and $f_{i+1 \bmod p+1}$ for $i = 0, \ldots, p$;

- the endpoints of each edge f_i, $i = 1, \ldots, p - 1$, are inside the closed curve C formed by the crossing points and the edge-pieces between these crossing points;

- only one endpoint of f_0, and only one endpoint of f_p are contained in C in the drawing $D_A(p)$;

- both endpoints of f_0 and f_p are contained in C in the drawing $D_B(p)$.

Note that the partial drawings $D_A(p)$ and $D_B(p)$ differ only in how the last edge is drawn with respect to the first one. Arroyo et al. [ABR20, Theorem 1.2] showed that such partial drawings are obstructions for pseudolinearity, that is, they cannot be part of any pseudoline arrangement. Therefore, neither of these partial drawings can be *completed* to a simple circular drawing, that is, the endpoints of the edges cannot be extended so that they lie on a circle which contains the drawing. We highlight this fact in the following lemma.

Lemma 7.2 ([ABR20]). *For a matching with $p + 1$ edges $f_0, f_1, \ldots, f_p$, neither the partial drawing $D_A(p)$ nor the partial drawing $D_B(p)$ can be completed to a simple circular drawing.*

Using this lemma, we can now prove the following statement.

(a) $D_A(p)$ (b) $D_B(p)$

Figure 7.6: The two types of partial drawings (for $p = 6$) and the closed curve C (light green) that they induce.

Lemma 7.3. *Let r be a region of the surface subdivision Ω, and let r' be its projection onto the plane. Then both r' and r are topological disks, that is homeomorphic to a disk (with the boundary). Moreover, the projection map is injective.*

Proof: Note that the boundary of r is formed by pieces of frame edges that were lifted on the surface S as described above and by additional corner-cuts as illustrated in Figure 7.5b in blue. This means that the boundary of r is a (closed) Jordan curve since we have only finitely many crossings in G. Then, we show that r does not include part of both a handle and its *undertunnel*, that is, the part of the surface over which the handle was built. This guarantees that the projection of r onto the plane is injective, and thus the boundary of r' is a Jordan curve. We will also show that r does not include holes. Then we can conclude that r' is homeomorphic to a disk using the Jordan–Schoenflies theorem, which says that for any closed Jordan curve there is a homeomorphism of the plane that maps the curve to the unit circle.

Suppose now, for a contradiction, that there are bundled crossings for which r contains both the handle and its undertunnel; see Figure 7.7a. Then there exists a non-intersecting Jordan arc $\gamma \subset r$ going over and under some of these handles. Consider the orthogonal projection γ' of γ on the disk of the drawing D (see Figure 7.7b) and notice that it self-intersects where it went over and under some handle in r. Choose the piece γ_1' of γ' separated by the self-intersection point X_Q, corresponding to some bundled crossing Q, such that γ_1' starts and ends in X_Q and no intersection point (except X_Q) is met twice when walking along γ_1' once; see Figure 7.7b.

Let P be the planarization of the projected drawing, and let P' be a copy of P without the edges that intersect γ_1'. Consider the edges of P' in the interior of the closed curve γ_1' (see Figure 7.7c) that can be *seen* from γ_1', that is, for each of such edges, we can draw a curve β from some point of γ_1' so that β does not intersect γ_1' and any other edge of P'. We call the edges of the drawing D that contain the edges of the planarization P' seen from γ_1' the *profile* of γ_1'; see Figure 7.7d. These edges form a partial drawing $D_A(p)$ for some $p > 0$; see Figure 7.6a. According to Lemma 7.2, however, such a partial drawing cannot be completed to a valid simple circular drawing; contradiction.

As for holes, it is easy to see that if r had a hole, the profile of any curve around this hole would yield a partial drawing $D_B(p)$ for some $p > 0$; see Figure 7.6b. Again, according to Lemma 7.2, such a partial drawing cannot be completed to a valid simple circular drawing; contradiction.

Note that since r' is a topological disk, its lifting r is also a topological disk. □

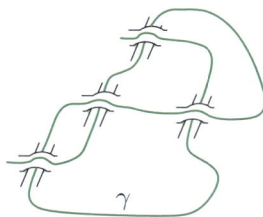

(a) a region r with handles and undertunnels corresponding to the same bundled crossings and a curve γ that goes over and under such handles

(b) projection of the region r and projection γ' of the curve γ on the drawing D in the plane; the curve γ_1' (light red) is a part of γ' that starts and ends in X_Q

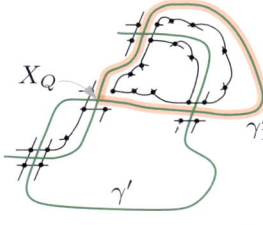

(c) the planarization P'

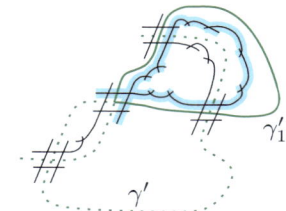

(d) the profile of the curve γ_1' (light blue); these edges form a partial drawing $D_A(p)$

Figure 7.7: Illustration to the proof of Lemma 7.3.

In particular, since our projection is injective, a drawing on r can be regarded as a drawing on r' and vice versa.

The next lemma concerning treewidth is a direct consequence of Lemma 7.3.

Lemma 7.4. *If a graph G admits a circular layout with k bundled crossings then its treewidth is at most $8k + 2$.*

Proof: If the graph G can be drawn in a circular layout with k bundled crossings then there exist at most $4k$ frame edges. According to Lemma 7.3, the removal of their endpoints breaks up the graph into outerplanar components. The treewidth of an outerplanar graph is at most two [Mit79]. Moreover, adding a vertex to a graph raises its treewidth by at most one. Thus, since deleting at most $8k$ frame vertices leaves behind an outerplanar graph, G has treewidth at most $8k + 2$. □

We now prove Theorem 7.2, which says that deciding whether $bc°(G) \leq k$ is FPT in k.

Proof of Theorem 7.2: We use Lemma 7.3 and extend the algorithm of Section 7.2.2.

Suppose that G has a circular drawing D with at most k bundled crossings. Then D contains a set F of (up to) $4k$ frame edges of these bundled crossings. As discussed before, F together with D defines a subdivided topological surface Ω containing a set R of regions. As in the case of one bundled crossing, each edge of G not in F is contained in exactly one such region,

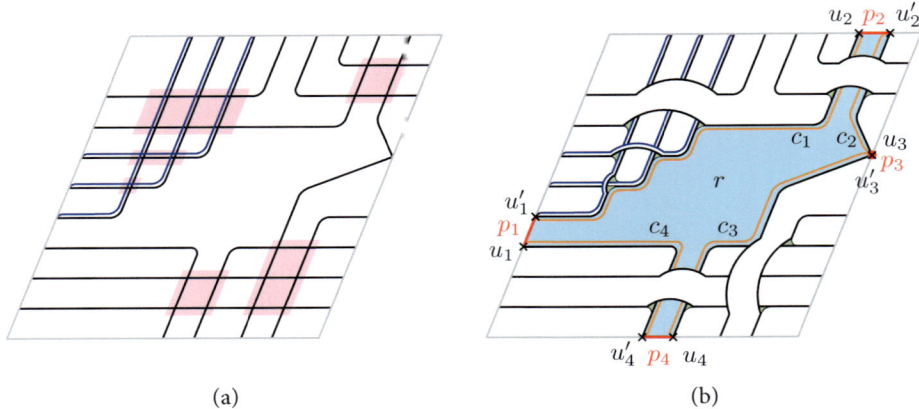

Figure 7.8: (a) A bundled drawing D with six bundled crossings (pink); parallel (blue) edges can be inserted to avoid degenerate bundled crossings; (b) the corresponding surface of genus 6; the components of the surface that are not regions are marked in green; the region r (light blue) has a boundary consisting of the arcs of the disk (red) and the arcs c_1, c_2, c_3, and c_4 (traced in orange).

and each vertex of G either is incident to an edge in F (in which case it belongs to at least two regions) or belongs to exactly one region.

Throughout the proof we refer the reader to Figure 7.8 for an example. By Lemma 7.3, each region r in R is a topological disk. Therefore, the graph G_r whose vertices lie on the boundary of r and whose edges lie in the interior of r is outerplane with respect to the given order of the vertices along the boundary. This boundary consists, in clockwise order, of arcs $p_1, \ldots, p_\alpha$ of the outer boundary of $\mathcal{S}$ (marked in red in Figure 7.8b) and Jordan arcs $c_1, \ldots, c_\alpha$ (traced in orange in Figure 7.8b), each of which connects two consecutive arcs of $\mathcal{S}$. For $i \in \{1, \ldots, \alpha\}$, let u_i and u_i' be the endpoints of p_i, in clockwise order. The arc p_i can degenerate to a single point; then $u_i = u_i'$; see Figure 7.9. So u_i' and u_{i+1} (where $u_{\alpha+1}$ is u_1) are the endpoints of c_i. No vertex of G_r lies in the interior of c .

We now describe G_r^*. First, we add a hub vertex h. Then, for each $i \in \{1, \ldots, \alpha\}$, if u_i' and u_{i+1} (where $u_{\alpha+1}$ is u_1) are not adjacent, we add an edge between them. If the arc p_i is non-degenerate, we add a boundary vertex b_i adjacent to all vertices on p_i (including u_i and u_i') and make h adjacent to u_i, b , and u_i'. Otherwise, we make h adjacent to $u_i = u_i'$ and identify b_i with u_i and u_i'. The reason for this identification is technical; it allows us to iterate over all (degenerate or non-degenerate) arcs and address their boundary vertices; see Section 7.2.4.

Observe that the resulting graph G_r^* is planar due to the special outerplanar drawing of G_r in r. Moreover, in every planar embedding of G_r^*, there is an outerplanar embedding of G_r where the cyclic order of the arcs c_i and the sets of vertices mapped to the p_i's match their cyclic order in r, implying that G_r fits into r. This is due to the fact that the simple cycle C' around h must be embedded planarly, with all of G_r inside (with the possible and easy-to-fix exceptions described in Section 7.2.2 concerning the cycle C there). Then the order of the

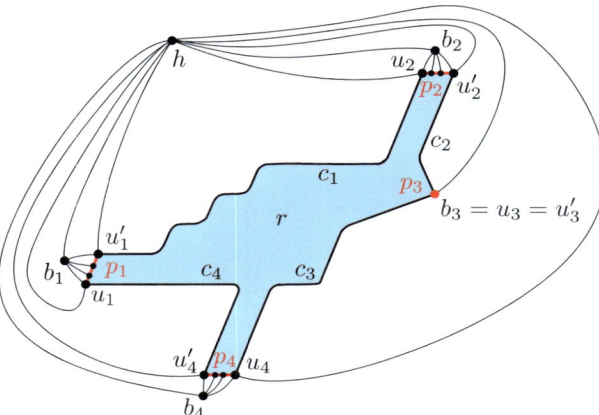

Figure 7.9: The augmented graph G_r^* for a complex region r. The arc p_3 is degenerate.

vertices in an outerplanar embedding of G_r is the order of the vertices incident to $b_1, \ldots, b_\alpha$ in a planar embedding of G_r^*. So the planarity of G_r^* guarantees that G_r fits into r as needed.

The reason why G has a circular drawing D with at most k bundled crossings is that there is a β-edge k-bundled crossing drawing D_F (of the graph formed by F), whose corresponding surface S consists of regions $r_1, \ldots, r_\gamma$ (note: $\gamma \leq 2\beta \leq 8k$) so that Properties 1–6 hold.

Our algorithm first checks that the treewidth of G is at most $8k + 2$. Recall that this can be done in linear time (FPT in k) [Bod96]. The algorithm then enumerates all possible simple drawings of at most $4k$ edges in the circle, i.e., at most $4k$ curves extending to infinity in both directions where each pair of curves cross at most once. The number of such drawings is proportional to k, and efficient enumeration has been done for the case when every pair of curves cross exactly once [Fel97]. For each drawing, it further enumerates the possible ways to form at most k bundled crossings so that every edge is a frame edge of at least one bundled crossing. Then, for each such bundled drawing D_F, we build an MSO$_2$ formula φ (see Section 7.2.4) to express Properties 1–6. Finally, since G has treewidth at most $8k + 2$, we can apply Courcelle's theorem on (G, φ). $\qquad\square$

7.2.4 MSO$_2$: Definitions and Our Formula for a Specific Layout of the Frame Edges

In this section we describe how to express the needed condition of our algorithm (as given by Properties 1–6) in MSO$_2$ logic (see Section 2.4).

We will describe our augmented graphs (from Property 6) as an MSO-transduction (see Section 2.4) and this will allow us (via [CE12, Theorem 7.10]) to have an MSO$_2$ formula to implicitly check the planarity of our augmented graphs inline within our (main) MSO$_2$ formula (where our formula is applied only to the graph prior to augmentation).

The formula for D_F.   We now construct an MSO_2 formula to express the following problem:

- Given a graph G with $V = V(G)$ and $E = E(G)$ and a simple circular drawing D_F with k bundled crossings so that $F = \{e_1, \ldots, e_\beta\}$ is the set of frame edges (and D_F has no other edges) and $V_F = \{v_1, \ldots, v_\xi\}$ is the set of frame vertices (and D_F has no other vertices);

- determine whether G has a simple circular drawing with k bundled crossings so that the frame edges and vertices occur as in D_F.

This is based on Properties 1–6 on page 94: we express them as MSO_2 formulas.

Properties 1 and 2 simply state that a set of elements is partitioned into a certain number of disjoint subsets. We use a formula stated by Bannister and Eppstein [BE18] to express this in MSO_2. For example, partitioning of a set E into $E_0, E_1, \ldots, E_\gamma$ disjoint subsets can be done in the following way.

$$\textsc{Partition}(E; E_0, \ldots, E_\gamma) = (\forall e \in E)\left[\left(\bigvee_{i=0}^{\gamma} e \in E_i\right) \wedge \left(\bigwedge_{i \neq j} \neg(e \in E_i \wedge e \in E_j)\right)\right].$$

We will additionally use the following formula to state that a vertex set V' is the set of endpoints of an edge set E':

$$\textsc{Incident}(V', E') = (\forall e \in E')\,(\forall v \in V(G))\left[I(e, v) \Leftrightarrow v \in V'\right].$$

We now turn to the properties more specific to our fixed drawing D_F of $\beta \leq 4k$ frame edges $F = \{e_1, e_2, \ldots, e_\beta\}$ whose endpoints are $V(F) = \{v_1, v_2, \ldots, v_\xi\}$, where $\xi \leq 2\beta$. As discussed in Section 7.2.1 and Lemma 7.3, this drawing induces a corresponding set of regions $r_1, \ldots, r_\gamma$.

Property 3 ensures that certain edges $E_0 = \{f_1, f_2, \ldots, f_\beta\}$ and their endpoints $V_0 = \{w_1, w_2, \ldots, w_\xi\}$ of the graph G induce a graph isomorphic to $(V(F), F)$. This can be modeled by the following formula.

$$\theta_3(V_0, E_0) = \left(\forall i, j \in \{1, 2, \ldots, \xi\}\right)$$
$$\left[\left((\exists f \in E_0)\, I(e, w_i) \wedge I(f, w_j)\right) \Leftrightarrow \left((\exists e \in F)\, I(e, v_i) \wedge I(e, v_j)\right)\right]$$

To express Properties 4 about the adjacencies of the frame vertices, we introduce the following piece of notation. For each vertex $v_i \in V(F)$ with $i \in \{1, 2, \ldots, \xi\}$, we denote by $\sigma(i)$ the set of indices of the regions incident to v_i in the drawing D_F. For example, in the case of one bundled crossing (see Figure 7.5), $\sigma(1) = \{1, 6\}$. Then Property 4 can be expressed in MSO_2 as follows:

$$\theta_4(V_0, E_0) = \left(\forall i \in \{1, 2, \ldots, \xi\}\right)(\forall e \in E)$$
$$\left[I(e, w_i) \Rightarrow \left[e \in E_0 \vee (\exists j \in \sigma(i))\left[e \in E_j\right]\right]\right].$$

Property 5 expresses that, for each non-frame vertex $v \in V_j$, all edges incident to v are contained in E_{i_j} (recall that i_j is the index of the region containing the set V_j of non-frame vertices and that E_{i_j} is the set of non-frame edges of this region):

$$\theta_5(V_1, \ldots, V_\eta) = \left(\forall j \in \{1, 2, \ldots, \eta\}\right)\left(\forall v \in V_j\right)\left(\forall e \in E\right)\left[I(e, v) \Rightarrow e \in E_{i_j}\right].$$

Finally, we turn to Property 6. First, note that testing planarity of a graph G can be expressed as follows where the formula for $\text{Minor}_H(G)$ does not need edge set quantification (i.e., it is in MSO) [CE12, Corollaries 1.14 and 1.15]:

$$\text{Planar}(G) = \neg\text{Minor}_{K_5}(G) \wedge \neg\text{Minor}_{K_{3,3}}(G).$$

Now, we describe the MSO-transduction τ_i of G to $G^*_{r_i}$ (for each region r_i; see Section 7.2.3) subject to the variables $w_1, \ldots, w_\xi, V_1, \ldots, V_\eta, f_1, \ldots, f_\beta, E_1, \ldots, E_\gamma$. Note that in our transduction, the input uses the format allowing for edge set quantification (i.e., where we have the objects $V \cup E$ and the binary incidence function I), but our output involves the format without edge set quantifications (i.e., where we have the objects V and the binary adjacency function adj). Let $j_1, \ldots, j_\zeta$ be the indices of the frame vertices incident to region r_i and suppose these are ordered cyclically as in D_F. Further, let $V_{l_1}, \ldots, V_{l_\alpha}$ be the sets corresponding to the arcs of the boundary of r_i (in order). With this notation, we can now set up the transduction τ_i which describes our graph $G^*_{r_i}$ in terms of our variables (note that in the statement of [CE12, Theorem 7.10] our variables are the *parameters*). Note that the symbols $h, b_1, b_2, \ldots, b_\alpha$ are new objects that are added in the construction (namely, the hub and boundary vertices of $G^*_{r_i}$). Further, let C be the cycle of the wheel, that is, $V(C) = \{v_{j_1}, \ldots, v_{j_\zeta}, b_1, \ldots, b_\alpha\}$. For each vertex $x \in V(C)$, let $N_C(x)$ be the set consisting of the two neighbors of x in C.

Now we can describe the transduction τ_i as follows.

$$V(G^*_{r_i}) := \{h\} \cup V(C) \cup \bigcup_{j=1}^{\alpha} V_{l_j}$$

$$
\begin{aligned}
adj_{G^*_{r_i}}(u,v) :=&(u \neq v) \wedge \\
&\Big(\big((\exists e \in E_i)\,(I(e,v) \wedge I(e,u))\big) \\
&\vee \big((h = u) \wedge (v \in V(C))\big) \vee \big((h = v) \wedge (u \in V(C))\big) \\
&\vee \Big(\bigvee_{j=1}^{\alpha} u = b_j \wedge v \in V_{l_j}\Big) \vee \Big(\bigvee_{j=1}^{\alpha} v = b_j \wedge u \in V_{l_j}\Big) \\
&\vee \big((u \in V(C)) \wedge (v \in N_C(u))\big) \\
&\vee \big((v \in V(C)) \wedge (u \in N_C(v))\big)\Big).
\end{aligned}
$$

With this transduction τ_i and the expression $\text{Planar}(G)$, we can now apply [CE12, Theorem 7.10] to obtain the MSO$_2$ formula ι_i which, when applied to G (together with our variables), allows us to express that $G^*_{r_i}$ is planar. Namely, by taking the conjunction of all of these ι_i, we obtain the needed MSO$_2$ formula ι (which can be applied to G and our variables) to express that all of the $G^*_{r_i}$'s are planar.

Now we construct the MSO_2 formula corresponding to Properties 1–6. The formula depends on the drawing D_F of the set of frame edges F.

$$\text{REALIZABLE}_{D_F}(G) \equiv$$
$$(\exists f_1, \dots, f_\beta, E_0, E_1, \dots, E_\gamma, w_1, w_2, \dots, w_\xi, V_0, V_1, \dots, V_\eta)$$
$$\Big[E_0 = \{f_1, \dots, f_\beta\} \wedge V_0 = \{w_1, w_2, \dots, w_\xi\}$$
$$\wedge \ \text{PARTITION}(E; E_0, E_1, \dots, E_\gamma)$$
$$\wedge \ \text{PARTITION}(V; V_0, V_1, \dots, V_\eta)$$
$$\wedge \ \text{INCIDENT}(V_0, E_0)$$
$$\wedge \ \theta_3(V_0, E_0) \wedge \theta_4(V_0, E_0) \wedge \theta_5(V_1, \dots, V_\eta)$$
$$\wedge \ \iota(f_1, \dots, f_\beta, E_1, \dots, E_\gamma, w_1, w_2, \dots, w_\xi, V_1, \dots, V_\eta) \Big].$$

7.3   Bundling a Drawing

We now establish the following parameterized results for bundling a given drawing.

Theorem 7.5. *Let G be a graph with n vertices and $m \geq n$ edges, and let D be a drawing of G.*

(a) *If D has c crossings, we can test whether $\text{bc}'(G, D) \leq k$ in $c^{O(\sqrt{k})} + O(m + c)$ time.*

(b) *If D is simple, we can test whether $\text{bc}(G, D) \leq k$ in $m^{O(\sqrt{k})} + O(m)$ time.*

(c) *If D is simple and circular, testing whether $\text{bc}^\circ(G, D) \leq k$ is FPT in k; it takes $2^{O(\sqrt{k}\log k)} + O(m)$ time.*

(d) *For a permutation π of $V(G)$, testing whether $\text{bc}^\circ(G, \pi) \leq k$ is FPT in k; it takes $2^{O(k^2)} + O(m)$ time.*

To prove this theorem, we will examine the number of combinatorially different bundled crossings we can make in our given fixed drawing D. Namely, let $\mathcal{B}(D)$ be the entire family of subsets of crossings in D such that each subset corresponds to a bundled crossing in D, that is, for each bundled crossing, the subset $\mathfrak{S}$ of the crossings in D contained in it is an element of $\mathcal{B}(D)$. We show that $|\mathcal{B}(D)| \in O(c^4)$ where c is the number of crossings in D; see Lemma 7.6. For the case when D is simple, we show that $|\mathcal{B}(D)| \in O(m^4)$ where m is the number of edges in D; see Lemma 7.5.

Note that each element of $\mathcal{B}(D)$ forms a connected subgraph in the planarization of the drawing. So, by finding k such connected subgraphs that are pairwise disjoint and together cover the crossings of D, we can bundle the drawing to have at most k crossings. Marx and Pilipczuk [MP15] studied exactly this type of disjoint covering problem. Their result is as follows.

Theorem 7.6 ([MP15, Theorem 1.3]). *Let G be a planar graph, let B be a family of connected vertex sets in G, let $C \subseteq V(G)$ be a set of vertices, and let k be an integer. In time $|B|^{O(\sqrt{k})}n^{O(1)}$, we can find a set S of at most k pairwise disjoint objects in B that maximizes the number of vertices of C in the union of the vertex sets in S.*

We use Theorem 7.6 and an algorithm of Alam et al. [AFP16] to prove Theorem 7.5.

Proof of Theorem 7.5: (a) Consider a drawing D of G with c crossings, and let $B = \mathcal{B}(D)$. By Lemma 7.6, $|B| \in O(c^4)$. Let D' be the plane graph obtained from D by creating a vertex at each crossing point in D (note that D' does not contain the vertices of G), and connecting two such vertices if they are consecutive along an edge in D. Clearly, each element in B forms a connected subgraph of D'. Thus, applying Theorem 7.6 with $G = D', B = \mathcal{B}(D), C = V(D')$ establishes (a).

(b) This follows as in (a). Namely, since D is simple, by Lemma 7.5, $|\mathcal{B}| \in O(m^4)$. This establishes (b).

(c) Alam et al. [AFP16] showed that testing $\mathrm{bc}^\circ(G, \pi) \le k$ can be kernelized down to an instance with at most $16k$ edges (or report that (G, π) is a no-instance) in $O(m)$ time. This also applies to a given simple circular drawing. Thus, by applying their kernelization and using (b) with $m \le 16k$, we establish (c).

(d) Here, we use (b) to improve the FPT algorithm of Alam et al. [AFP16] for determining whether $\mathrm{bc}^\circ(G, \pi) \le k$. (Their algorithm runs in time $k^{O(k^2)} + O(m)$ and proceeds in three stages). First, it applies a kernelization step to obtain a graph with at most $16k$ edges in $O(m)$ time; second, it enumerates all possible $O\left(2^{0.657k^2}\right)$ weak pseudoline arrangements of $16k$ pseudolines [Fel97]; and third, for each such weak pseudoline arrangement it partitions the crossings into the minimum number of bundled crossings by exhaustive search in time $O\left(k^{128k^2}\right)$. For this last step, we apply (b) instead (now with $m \le 16k$), leading to $(16k)^{O(\sqrt{k})} = 2^{O(\sqrt{k} \log k)}$ time, and $2^{O(k^2)} + O(m)$ time in total. $\qquad\square$

Note that since the number of crossings in a non-simple drawing is not bounded by a function of the number of edges m in the drawing, we do not obtain an analogous result for the circular layout as in Theorem 7.5(c) by using the kernelization technique of Alam et al. [AFP16]. On the other hand, we present a $2^{O(m)}$-time algorithm for not necessarily simple drawings in circular layouts in the context of storyline visualization; see Section 7.4.

We now discuss the size of the family $\mathcal{B}(D)$ for some drawing D. Note that each bundled crossing in D involves two pairs of frame arcs, and, conversely, two pairs of frame arcs can determine at most one bundled crossing. We show that if D is simple, then this is also true for frame edges, that is, two pairs of frame edges can determine at most one bundled crossing; see Lemma 7.5. This allows us to bound the number of distinct bundled crossings by the number of edges from above, and thus, the size of the family $\mathcal{B}(D)$.

Lemma 7.5. *Let D be a simple drawing D with m edges. Each bundled crossing determines at most two pairs of frame edges, and, conversely, two pairs of frame edges can determine at most one bundled crossing. In particular, $|\mathcal{B}(D)| \le m^4$.*

Proof: Consider two bundles that form a given bundled crossing. Each bundle has at most two not necessarily distinct frame edges. For the reverse direction, consider a bundled crossing B and let (e_2, e_4), (e_1, e_3) be the two pairs of frame edges each corresponding to a bundle. Let c_{ij}, for $i = 1, 3$, $j = 2, 4$ be the frame crossing of e_i and e_j (if it exists). There are three cases how these two pairs can determine a bundled crossing: (a) $e_1 = e_3$ and $e_2 = e_4$: then B is a single crossing, clearly there cannot be another bundled crossing determined by the same pairs; (b) $e_1 = e_3$ and $e_2 \neq e_4$ (the case where $e_1 \neq e_3$ and $e_2 = e_4$ is symmetric): then B consists of all crossings on e_1 between c_{12} and c_{14}, since e_1 and e_2 can cross e_1 at most once, there cannot be another bundled crossing; or (c) the edges are pairwise different, then all the crossings c_{ij} exist and are distinct and any other bundled crossing would imply that for some fixed i and j the crossings c_{ij} occurred twice, which is impossible in a simple drawing.   □

In the case of a not necessarily simple drawing D, the number c of crossings cannot be bounded in terms of the number m of edges. But if we define the size of an instance in terms of the number of crossings, it is easy to see that $\mathcal{B}(D)$ is of size polynomial in c.

Lemma 7.6. *For any not necessarily simple drawing D with c crossings $|\mathcal{B}(D)| \in O(c^4)$.*

Proof: Every bundled crossing can be determined by two pairs of frame arcs. Since each crossing can be incident to at most four arcs, four crossings can determine at most 4^4 different bundled crossings. Therefore, the total number of bundled crossings is polynomially bounded by the number of crossings, namely $|\mathcal{B}(D)| \leq 16c^4$.   □

7.4 Bundling Storyline Visualisations Is FPT

For our purposes, a storyline drawing D is a set of m x-monotone curves. Such curves cannot self-intersect, but a pair of curves is allowed to intersect each other multiple times; we only forbid the existence of digonal faces, that is, two curves intersecting each other twice in a row. Finally, we assume here that all curves start on distinct points of a vertical line v_{left} and end on distinct points of a vertical line v_{right}. This is common in storyline visualizations, but this restriction can be dropped with additional care. We prove the following.

Theorem 7.7. *Given a storyline drawing D with m characters and c crossings, $\mathrm{bc}^s(D)$ can be computed in $O(\varphi^{2m} mc) \subset O(2.62^m mc)$ time, where φ is the golden ratio. This runtime is fixed parameter tractable in m. An optimal bundling can be constructed in the same time.*

Recall that $I(D)$ is the set of crossings. Each curve, being x-monotone, gives a left-to-right order of its incident crossings. These orders give a partial order on $I(D)$. Let π be an arbitrary linear extension of these partial orders, which can be found in polynomial time given D. Then we subdivide D into columns according to π: see Definition 7.2 and Figure 7.10. We call a face of this subdivision a *cell*. In this section we define a way to label the cells to describe any bundling of the drawing. The algorithm for Theorem 7.7 is based on dynamic programming over such labelings.

Definition 7.2. A subdivision $\mathcal{S}(D)$ consists of the drawing D together with: horizontal lines h_{top} and h_{bot} above and below all curves; vertical lines v_{left} and v_{right} going through the left and right endpoints of the curves, respectively; and a set of c y-monotone curves with the following properties:

- for each crossing X in $I(D)$, there is a unique curve going through X,
- each curve crosses h_{top}, h_{bot} and all curves of D,

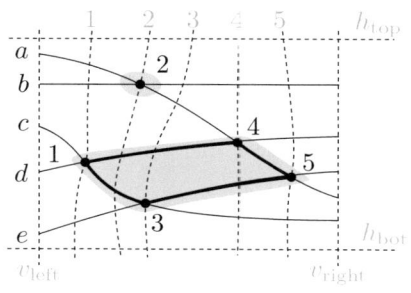 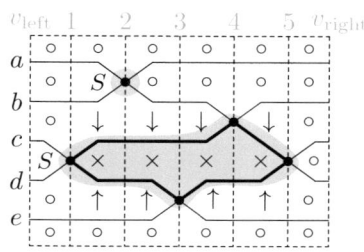

(a) bundling of D with order π (b) corresponding real labeling of $\mathcal{S}(D)$

Figure 7.10: Bundling of a storyline drawing: curves added in $\mathcal{S}(D)$ are dashed and a bundling of the crossings is indicated in gray. Note the degenerate bundled crossing at crossing 2.

- the curves do not intersect each other, and they are totally ordered from left to right according to π.

See Figure 7.10a for an example of a drawing D with its subdivision $\mathcal{S}(D)$. Let $\mathcal{C}$ be the set of bounded faces of $\mathcal{S}(D)$; we call the elements of $\mathcal{C}$ the cells in order to distinguish them from the faces of D. A drawing of $\mathcal{S}(D)$ in Figure 7.10b helps to understand its structure. (This drawing is stretched similarly to a *wiring diagram* [FG18].) Note that the subdivision consists of $|I(D)| + 1$ *columns*, each with $m + 1$ cells, and all cells are either triangular or quadrangular: there are triangles to the left and right of each intersection, and all other cells are quadrangles. The cells in a column are numbered from top to bottom, starting at 1. These numbers are their *row* numbers.

We use the set $\mathcal{L} = \{\times, S, \circ, \downarrow, \uparrow, \updownarrow\}$ of labels. In a fixed bundling of D, each cell satisfies exactly one of the following conditions.

$\times$ This cell is inside a bundled crossing. (This can only happen if the cell is part of a quadrangular face.)

For cells not inside a bundled crossing, there are five options.

S   This cell is directly left of the π-earliest crossing of a bundled crossing: it "starts" a bundled crossing.

$\circ$ This cell does not touch the boundary of a bundled crossing, except possibly in a point.

$\downarrow$ Only the lower boundary of this cell bounds a bundled crossing.

$\uparrow$ Only the upper boundary of this cell bounds a bundled crossing.

$\updownarrow$ Both the upper and lower boundary of this cell bound a bundled crossing.

We call a function from the set $\mathcal{C}$ of cells to the set $\mathcal{L}$ of labels a *labeling*. A labeling is called *real* if there exists a bundling of D where each cell satisfies the condition of its label. We now observe four necessary properties of real labelings. Afterward, we prove that they are sufficient.

Crossing property Any crossing must be part of a bundled crossing (though possibly a degenerate one). Consider the six cells surrounding a crossing in $S(D)$. If the crossing lies in the interior of a bundled crossing, all six cells are labeled × in the real labeling. Otherwise it lies on the boundary of a bundled crossing. Enumeration reveals the finite set of ways to label these cells that can possibly be real: see Figure 7.11. Any other way to label the six cells around the crossing directly contradicts the conditions for the labels.

Column property Note that in a real labeling, a column of $S(D)$ can have at most one cell labeled 'S', since by construction only one cell per column is left of a crossing. Now consider the sequence of labels encountered top to bottom in a column, for example $[\circ, \circ, \downarrow, \times, \uparrow, \circ]$ in the third column of Figure 7.10b. For any real labeling, this sequence describes being inside (×) and outside (∘) of bundled crossings, with $\downarrow$, $\uparrow$ and $\updownarrow$ marking the transitions, and possibly the label 'S' in place of ∘ in one particular cell (left of the crossing). Such sequences without 'S' are walks through the directed graph in Figure 7.12; the Column property says that precisely these sequences are allowed, with the addition of also allowing 'S' in the appropriate cell (and only there).

Row property In any real labeling, horizontally adjacent cells have the same label unless they share a crossing on their boundary (and in that case the change is governed by the Crossing property), where ∘ and 'S' are considered the same: the labels ∘ and 'S' both describe a cell that does not touch the boundary of a bundled crossing. Such pairs of horizontally adjacent cells must have the same label, because they have the same incidences to any bundled crossings: these incidences can only change at crossings.

Consider for example the sequence of labels encountered left to right in the third row for Figure 7.10b: $[\circ, \downarrow, \downarrow, \downarrow, \downarrow, \circ]$. For the Row property, the third and fourth labels must be the same; the other pairs are exempt due to sharing a crossing. The 'S' in the second row is allowed to the right of ∘ since the two labels are considered equal for the Row property.

Quadrangle property In any real labeling, all faces inside a bundled crossing are quadrangles. Therefore, only cells contained in a quadrangular face of D can have the label ×.

These properties are necessary for real labelings (as argued above) and, once we fix the leftmost and rightmost columns of the labeling, they are also sufficient.

Lemma 7.7. *A labeling $C \to \mathcal{L}$ is real if and only if:*

1. *the first column contains 'S' left of its crossing and ∘ everywhere else,*
2. *the last column contains ∘ everywhere, and*
3. *the Crossing, Column, Row, and Quadrangle properties hold everywhere.*

The number of bundled crossings in the corresponding bundling equals the number of cells with label 'S'.

Proof: First observe that in any real labeling, the first column consists of all ∘ except the label 'S' in the unique cell adjacent to a crossing: none of the cells bound (or are in) a bundled crossing except that the 'S' cell necessarily touches one in a point.

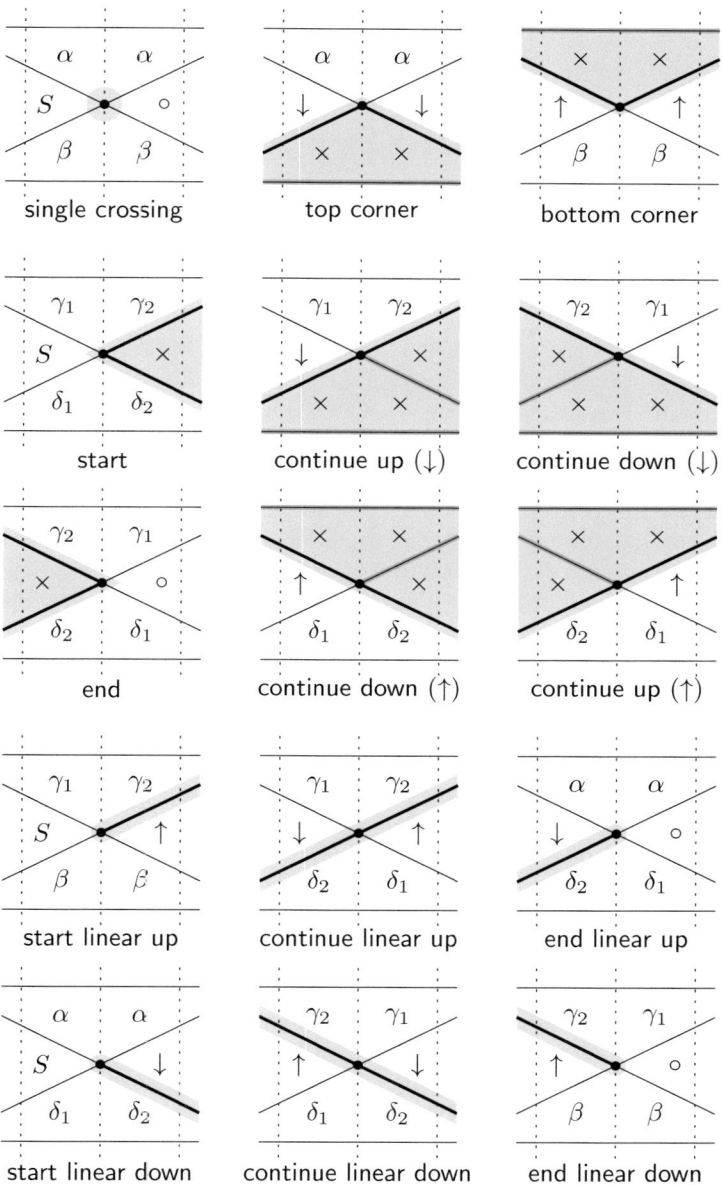

Figure 7.11: All possible configurations of labels around a crossing on the boundary of a bundled crossing, where the Greek variables may be substituted as follows: $\alpha \in \{\circ, \uparrow\}$, $\beta \in \{\circ, \downarrow\}$, $\gamma_1 \gamma_2 \in \{\circ \downarrow$ $, \uparrow \updownarrow\}$, $\delta_1 \delta_2 \in \{\circ \uparrow, \downarrow \updownarrow\}$. Multiple occurrences of the same variable within one configuration must be substituted consistently, so for example the two α in the top left configuration must both be $\circ$ or both be $\uparrow$. (If the crossing is in the interior of a bounded crossing, all six cells must be $\times$.)

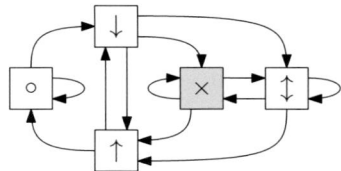

Figure 7.12: Directed graph for the Column property, reading a column top to bottom; starting nodes are ∘ and ↓. (Further legal columns can be obtained by replacing one occurrence of ∘ by 'S'.) Only the gray state corresponds to cells inside of a bundled crossing.

Similarly in any real labeling the last column cannot have cells that are contained in a bundled crossing or adjacent to a crossing on their right, so they are all labeled ∘. As argued above, the Crossing, Column, Row, and Quadrangle properties hold everywhere in any real labelling. This establishes one direction of the lemma.

We now show that if a labelling satisfies the three conditions of the lemma, then it is real. Consider a connected component of cells labeled ×. Based on the Row and Crossing properties, the surrounding cells are correctly labeled with arrows and an 'S' label. Call such a connected component a *blob*. We show that each blob is indeed a bundled crossing by Definition 7.1. (Note that blobs are nondegenerate bundled crossings; we handle degenerate bundled crossings later.)

Let B be a blob. Call a crossing on the boundary of B a *convex corner* if no curve in this crossing goes into the interior of the blob; a *side* if one curve goes into the interior, or; a *reflex corner* if two curves go into the interior. Notice that the "start", "top corner", "bottom corner", and "end" configurations of the Crossing property represent convex corners, and the other configurations with × represent sides. Therefore, a blob cannot have reflex corners. Thus, a blob is a topological disk. Moreover, the first column of $\mathcal{S}(D)$ contains no × labels, so a blob has a unique "start" configuration.

Now we trace the boundary of B, starting at its π-earliest crossing X, and we will find the four frame arcs. Start from X and follow the boundary along one of the curves and call this frame arc e_1. We switch to the next frame arc whenever we encounter a convex corner. This process is repeated until we get back to X, which must happen because the blob is a topological disk and bounded on the right by the last column (which does not contain any ×). Let $e_1, e_2, \ldots, e_k$ be the frame arcs encountered. We only switched from a frame arc to the next at convex corners, and according to the Quadrangle property all faces in the blob are quadrangles. Then $k = 4$, since that is the only way to close a loop around quadrangles using only convex corners. In fact, since all faces in the blob are quadrangles, the crossings inside B form a grid and therefore B is a bundled crossing.

Consider a cell labeled 'S' that does not precede a blob. If the "single crossing" configuration of the Crossing property applies, this correctly describes a singleton bundled crossing. Otherwise, a "start linear up" or "start linear down" configuration must apply (Crossing property). Because of the "continue linear" configurations and the Row property, this must propagate and can only end in an "end linear up" or "end linear down" configuration. This correctly describes a linear bundled crossing.

Therefore, a labeling that starts correctly in the first column, where the properties hold everywhere, and that arrives correctly in the final column, is real. □

In preparation for the runtime bound of Theorem 7.7, we now bound the number of ways to label a column that are consistent with the Column property.

Lemma 7.8. *The number of length-n strings over $\mathcal{L}$ that are consistent with the Column property is $O(\varphi^{2n}) \subset O(2.62^n)$, where φ is the golden ratio. The strings can be enumerated with linear-time overhead.*

Proof: Enumerating the walks in the graph from Figure 7.12, with the additional option of having 'S' in place of ∘ in one particular cell, can be achieved in depth-first fashion using a stack. The optional 'S' at most doubles the number of accepted strings, so we ignore it for the asymptotic analysis and consider only the walks in the graph. We also ignore that there are two starting nodes (∘ and ↓), since again this only involves a factor two.

Now consider the adjacency matrix A of the graph; here the nodes are given in the order ∘, ↓, ↑, ×, ↕.

$$A = \begin{pmatrix} 1 & 1 & 0 & 0 & 0 \\ 0 & 0 & 1 & 1 & 1 \\ 1 & 1 & 0 & 0 & 0 \\ 0 & 0 & 1 & 1 & 1 \\ 0 & 0 & 1 & 1 & 1 \end{pmatrix}$$

The characteristic polynomial $\det(A - \lambda I)$ of the matrix A is $-\lambda^5 + 3\lambda^4 - \lambda^3$. Its roots are 0 of multiplicity three, $\frac{1}{2}(3 + \sqrt{5}) = \varphi^2$, and φ^{-2}, where φ is the golden ratio. Since the root with the largest absolute value is $\varphi^2 > 1$ and it has multiplicity one, the number of walks of length n is $O(\varphi^{2n})$. The lemma follows. (See, for example, Ardila's treatment [Ard15] of algebraic methods for counting walks.) □

Proof of Theorem 7.7: We use dynamic programming, moving from left to right by column of $\mathcal{S}(D)$: with $L \in \mathcal{L}^{m+1}$ and i a column, let $f(L, i)$ be the minimum number of 'S' labels in any labeling of the columns up to column i, ending with the labels L for column i. By Lemma 7.8, there are only $O(\varphi^{2m})$ values of L that satisfy the Column property and they can be enumerated with linear overhead. Each individual $f(L, i)$ can be computed with a constant number of lookups of $f(\cdot, i-1)$: by the Row property only the three rows adjacent to the crossing between the columns can change and the Crossing property gives a finite set of options for how they can change.

See pseudocode below, where $F(L, i)$ is used to store and look up values of $f(L, i)$; we use the convention that accessing $F(L, i)$ returns ∞ if that value has not been stored yet. The values can be accessed in $O(m)$ time by storing them in a prefix tree (indexed by L) per column. This leads to a total runtime of $O(\varphi^{2m} mc)$. □

If desired, the bundling itself can be read from F. In that case, the algorithm uses $O(\varphi^{2m} mc)$ space: the c prefix trees each store $O(\varphi^{2m})$ items and have height $m + 1$. If only the bundled crossing *number* is required, space usage can be improved to $O(\varphi^{2m} m)$ by storing only two columns at a time.

Input: Drawing D.
Output: Bundled crossing number $\mathrm{bc}^s(D)$.
$C \leftarrow$ Columns of $\mathcal{S}(D)$ numbered 0 to c
$L_{\mathrm{start}} \leftarrow$ Entire column is $\circ$, except 'S' left of the crossing of $C[0]$
$F(L_{\mathrm{start}}, 0) \leftarrow 1$ // Dynamic programming data structure
for $i \leftarrow 1$ **to** c **do**   // $O(c)$ times
    foreach $L \in$ Enum-Single-Columns$(C[i])$ **do**   // $O(\varphi^{2m})$ times
        foreach $L' \in$ Enum-Valid-Predecessors$(C[i-1], C[i], L)$ **do**
             $F(L, i) \leftarrow \min\{F(L, i), F(L', i-1)\}$ // $\curvearrowright O(1)$ times
      if L contains 'S' **then** $F(L, i) \leftarrow F(L, i) + 1$

$L_{\mathrm{end}} \leftarrow$ Entire column is $\circ$
return $F(L_{\mathrm{end}}, c)$

Subroutine: Enum-Single-Columns(c)
Input: A single column c of $\mathcal{S}(D)$.
Output: Enumerates all ways to label the column according to the Column property.

Subroutine: Enum-Valid-Predecessors$(c_{\mathrm{pred}}, c_{\mathrm{curr}}, L)$
Input: Adjacent columns c_{pred} and c_{curr} of $\mathcal{S}(D)$, labeling L for c_{curr}.
Output: Enumerates all ways to label c_{pred} according to the four properties, given that c_{curr} is labeled L.

7.5 Discussions and Open Problems

Given our new FPT algorithm for simple circular layouts, it would be interesting to improve its runtime and to investigate whether a similar result can be obtained for general simple layouts. A starting point could be the FPT algorithm of Kawarabayashi et al. [KR07] for computing the usual crossing number of a graph. We also conjecture that it is NP-hard to compute $\mathrm{bc}^\circ(G, \pi)$, given a graph G and a vertex order π. It seems plausible to reduce from SortingByTranspositions, but it is difficult to keep the resulting drawings simple.

We remind the reader of the open problem posed by Alam et al. [AFP16] and Fink et al. [FHSV16] concerning the computational complexity of $\mathrm{bc}^\circ(G)$.

Conclusion

In this book, we have investigated two areas of Beyond Planarity: crossing optimization in circular-arc drawings and crossing optimization in circular layouts. For crossing optimization in circular-arc drawings we have first considered a way to draw graphs without crossings while at the same time optimizing quality measures that affect the readability of a drawing like the visual complexity. Then we have analyzed properties of orthogonal circles as well as properties of arrangements of such circles. Lastly we have introduced and studied a new class of beyond-planar graphs, namely, graphs that can be drawn with circular-arc edges and right-angle crossings.

For crossing optimization in circular layouts we have considered several classes of beyond-planar graphs with forbidden edge crossing patterns, that is, graphs that have circular layouts where each edge is crossed at most k times and graphs that have circular layouts where there are no k pairwise crossing edges. We have studied their properties and designed recognition algorithms. Finally we have considered an effective way to reduce clutter in a graph drawing that has (many) crossings, that is, to group edges that travel in parallel into bundles. Each edge can participate in many such bundles. Any crossing in this bundled graph occurs between two bundles, i.e., as a bundled crossing. We have considered different variants of bundled crossing minimization, in particular, in circular layouts.

We will now recap the main results from individual chapters of this book and state some selected problems that are still open.

We have introduced (in Chapter 3) the spherical cover number $\sigma_d^l(G)$ of a graph G, that is, the minimum number of l-dimensional spheres in $\mathbb{R}^d$ such that G has a crossing-free circular-arc drawing that is contained in the union of these spheres. We have compared the spherical cover number to some other graph parameters, in particular, to its close relative the affine cover number $\rho_d^l(G)$, the minimum number of l-dimensional affine subspaces in $\mathbb{R}^d$ that together cover a crossing-free straight-line drawing of G. Obviously $\sigma_d^l(G) \leq \rho_d^l(G)$. On the other hand we have seen that $\rho_2^1(G) \in O(\sigma_2^1(G)^2)$. Furthermore, we have already seen that $\sigma_3^2(K_n)$ grows asymptotically more slowly than $\rho_3^2(K_n)$. Families of graphs where there is an asymptotic difference between the two cover numbers would be particularly interesting for $l = 1$ and $d = 2$ (recall that then we consider only planar graphs).

Open Problem 1. Is there a family of planar graphs where σ_2^1 grows asymptotically more slowly than ρ_2^1?

As a first step to study circular-arc drawings with crossings we have analyzed properties of orthogonal circles (in Chapter 4), in particular, properties of arrangements of such circles. We have shown that every arrangement of n orthogonal circles has at most $14n$ intersection points and $15n + 2$ faces. In addition we have provided upper bounds for the maximum number of faces with small degree, that is, for digonal faces and triangular faces. Namely, every arrangement of n orthogonal circles has at most $2n$ digonal faces and $4n$ triangular faces. For the lower bounds we have only obtained arrangements with $2n - 2$ digonal, $3n - 3$ triangular, and $4(n - 3)$ quadrangular faces. Note that the lower and upper bounds are tight only for the number of digonal faces. Thus, an immediate open question is to improve the bounds that are not tight, in particular, the lower bounds for the triangular and quadrangular faces.

Open Problem 2. Is there a better upper bound on the number of faces in an arrangement of orthogonal circles with n circles than $15n + 2$?

Open Problem 3. Are there arrangements of orthogonal circles with n circles and more than $3n - 3$ triangular or $4(n - 3)$ quadrangular faces?

We have introduced a new class of beyond-planar graphs, called arc-RAC graphs, that is, graphs that can be drawn with circular arcs and right-angle crossings; see Chapter 5. The class of arc-RAC graphs generalizes the known class of RAC graphs, where the crossings also occur at right angles only but the edges are drawn straight-line. We have provided a linear upper bound on the maximum edge density of the arc-RAC graphs of $14n - 12$. In addition we have constructed a family of n-vertex arc-RAC graphs with $4.5n - O(\sqrt{n})$ edges, which shows that arc-RAC graphs is a proper superclass of RAC graphs.

Open Problem 4. Can we improve the upper or the lower bound on the maximum edge density of arc-RAC graphs?

The relationship of the families of RAC and RAC_1 graphs (that is, graphs with drawings where edges are poly-lines with only one bend and the crossings are still right-angle only) to k-planar graphs (that is, graphs that can be drawn with at most k crossings per edge) is well understood. In particular, it is known that there are 1-planar graphs that are not RAC and that every 1-planar graph is RAC_1. Thus, it is intriguing to find out what is the relationship between arc-RAC and 1-planar graphs.

Open Problem 5. Is there a 1-planar graph which is not arc-RAC?

After studying circular-arc drawings we turned to crossing optimization in circular layouts (in Chapter 6), in particular, we have considered the following classes of beyond-planar graphs with forbidden edge crossing patterns: outer k-planar graphs, that is, the graphs that admit a circular layout where each edge is crossed by at most k other edges; and outer k-quasi-planar graphs, that is, the graphs that admit a circular layout where no k edges cross pairwise. For outer k-planar graphs we have given bounds with respect to k on several graph parameters for these graphs, in particular, the balanced separation number and the chromatic number. For each fixed k, we have shown, using small balanced separators, how to test outer k-planarity in quasi-polynomial time. For outer k-quasi-planar graphs we have compered them to other graph classes, in particular, to planar graphs.

We have considered restrictions of outer k-planar and outer k-quasi-planar drawings to full drawings (where no crossing appears on the boundary), and to closed drawings (where the vertex sequence on the boundary is a cycle in the graph) and provided linear recognition algorithms based on extended monadic second-order logic and Courcelle's Theorem (see Section 2.4 for definitions).

One of the open problems that stems from this chapter is on the relation of quasi-planar graphs and planar graphs. Every planar graph is outer 5-quasi-planar because planar graphs have page number 4 [Yan89] (planar graphs that require 4 pages have also been discovered recently [Yan20, BKK⁺20]). There are also planar graphs that are not outer 3-quasi-planar.

Open Problem 6. Is every planar graph outer 4-quasi-planar?

Another open problem arises from the closeness of outer k-quasi-planar with the class of circle graphs, that is, the class of intersection graphs of chords of a circle (also known as chord diagrams). An outer k-quasi-planar drawing D of a graph G can be seen as a chord diagram and as such provides a corresponding graph H where each adjacency between vertices corresponds to a crossing between edges of our drawing. Thus, k-coloring H corresponds to partitioning D into a k-page book embedding of G, that is, a drawing in k half-planes, called pages, that all intersect at the same line where the vertices are drawn. A q-clique in H with $q < k$ corresponds to some q pairwise crossing edges in the corresponding outer k-quasi-planar drawing D. Ageev [Age96] showed that any triangle-free circle graph has chromatic number at most 5. Therefore, any outer quasi-planar graph G has page number at most 5. This immediately gives rise to the following question.

Open Problem 7. Is there an outer quasi-planar graph with page number 5?

Finally, we have considered bundled crossing minimization in Chapter 7. A graph is given and the goal is to find a bundled drawing with at most k bundled crossings. We have shown that the problem is NP-hard when we require a simple drawing, resolving an open problem by Fink et al. [FHSV16]. Our main result is an FPT algorithm (in k) for simple (that is, no edge self-intersects and no two edges intersect twice) circular layouts, which answers an open question by Alam et al. [AFP16]. The algorithm is based on extended monadic second-order logic and Courcelle's Theorem (see Section 2.4 for definitions). Even though the runtime of testing a graph property expressed via a monadic second-order logic formula of bounded size is linear in the size of the graph (if the graph has bounded treewidth) it is notoriously large with respect to the parameters (that is, the treewidth and the size of the formula). Thus, a faster algorithm is desirable.

Open Problem 8. Is there a faster FPT algorithm that for a given graph G and a natural number k computes a simple circular layout with k bundled crossings if one exists?

So far we still do not know whether the aforementioned problem is NP-hard.

Open Problem 9 (Alam et al. [AFP16] and Fink et al. [FHSV16])**.** What is the complexity of deciding whether for a given graph G and a natural number k there is a simple circular layout with k bundled crossings?

Acknowledgments

My first acknowledgments go to my supervisor Alexander "Sascha" Wolff. I met him the first time at a Computational Geometry block course as a master student and I was inspired by his lectures, the slides, and the way how he presented the material. When I started at Sascha's chair he gave me immense support researchwise and not only. Sascha gave me lots of interesting research ideas and helped me to persevere with our research plan. Sascha also provided a research-friendly diverse and multi-sided environment at his chair which helped me to develop as a researcher. I want to thank Sascha for making research fun and for giving his students a possibility to travel around the world to different conferences and workshops.

I would also like to thank Steve Chaplick and Thomas van Dijk for a helpful cooperation and mentoring, in particular, for their readiness to answer my questions whenever I knocked at their doors. Further, I would like to thank my (former) colleagues for interesting conversations and for having fun when traveling together to conferences or just on our way to Mensa.

I would also like to especially thank Alexander "Sasha" Ravsky for his help, mentoring, and inspiring conversations that we had during our evening strolls.

Further, I would like to express my gratitude to co-authors and researchers from other universities that I had a pleasure to work with : Giuseppe Liotta, André Schulz, Herny Förster, and Ji-won Park. Also I would like to acknowledge Alon Efrat and Bruno Courcelle for their advice.

I am especially grateful to my parents for inspiring in me the love for science from my early age and my older brother and sister who I can always look up to.

Finally, I would like to thank the supervisor of my master thesis, Prof. Dr. Roman Chapko, for helping to obtain the DAAD scholarship for my PhD studies. The financial support of DAAD and their coordination is greatly acknowledged.

Bibliography

[AAA⁺12]   Oswin Aichholzer, Wolfgang Aigner, Franz Aurenhammer, Kateřina Čech Do-biášová, Bert Jüttler, and Günter Rote. Triangulations with circular arcs. In Marc van Kreveld and Bettina Speckmann, editors, *Proc. Graph Drawing (GD'11)*, volume 7034 of *Lecture Notes Comput. Sci.*, pages 296–307. Springer, 2012. [see page 2]

[AAS03]   Pankaj K. Agarwal, Boris Aronov, and Micha Sharir. On the complexity of many faces in arrangements of pseudo-segments and circles. In Boris Aronov, Saugata Basu, János Pach, and Micha Sharir, editors, *Discrete and Computational Geometry: The Goodman–Pollack Festschrift*, pages 1–24. Springer, 2003. [see page 39]

[ABB⁺16]   Christopher Auer, Christian Bachmaier, Franz J. Brandenburg, Andreas Gleißner, Kathrin Hanauer, Daniel Neuwirth, and Josef Reislhuber. Outer 1-planar graphs. *Algorithmica*, 74(4):1293–1320, 2016. [see page 71]

[ABB⁺17]   Patrizio Angelini, Michael A. Bekos, Franz J. Brandenburg, Giordano Da Lozzo, Giuseppe Di Battista, Walter Didimo, Giuseppe Liotta, Fabrizio Montecchiani, and Ignaz Rutter. On the relationship between k-planar and k-quasi planar graphs. In Hans Bodlaender and Gerhard Woeginger, editors, *WG 2017*, volume 10520 of *Lecture Notes Comput. Sci.*, pages 59–74, 2017. [see page 72]

[ABFK18]   Patrizio Angelini, Michael A. Bekos, Henry Förster, and Michael Kaufmann. On RAC drawings of graphs with one bend per edge. In Therese Biedl and Andreas Kerren, editors, *Proc. Graph Drawing & Network Vis. (GD'18)*, volume 11282 of *LNCS*, pages 123–136. Springer, 2018. [see page 55]

[ABR20]   Alan Arroyo, Julien Bensmail, and R. Bruce Richter. Extending drawings of graphs to arrangements of pseudolines. In Sergio Cabello and Danny Z. Chen, editors, *SoCG*, 2020. To appear. [see page 95]

[Ack09]   Eyal Ackerman. On the maximum number of edges in topological graphs with no four pairwise crossing edges. *Discrete Comput. Geom.*, 41(3):365–375, 2009. [see pages 55, 56, 57, 58, 59, and 71]

[Ack19]   Eyal Ackerman. On topological graphs with at most four crossings per edge. *Computational Geometry*, 85:101574, 2019. [see pages 71 and 74]

[ACNS82]   M. Ajtai, V. Chvátal, M.M. Newborn, and E. Szemerédi. Crossing-Free Sub-graphs. In Peter L. Hammer, Alexander Rosa, Gert Sabidussi, and Jean Turgeon, editors, *Theory and Practice of Combinatorics*, volume 60 of *North-Holland Mathematics Studies*, pages 9 – 12. North-Holland, 1982. [see pages 2 and 85]

Bibliography

[AEH80]   Jin Akiyama, Geoffrey Exoo, and Frank Harary. Covering and packing ingraphs III: Cyclic and acyclic invariants. *Math. Slovaca*, 30:405–417, 1980. [see page 33]

[AFK⁺12]   Karin Arikushi, Radoslav Fulek, Baláazs Keszegh, Filip Morić, and Csaba D. Tóth. Graphs that admit right angle crossing drawings. *Comput. Geom.*, 45(4):169–177, 2012. [see pages 55, 56, 64, 66, and 67]

[AFP16]   Md. Jawaherul Alam, Martin Fink, and Sergey Pupyrev. The bundled crossing number. In Yifan Hu and Martin Nöllenburg, editors, *GD*, volume 9801 of *Lecture Notes Comput. Sci.*, pages 399–412. Springer-Verlag, 2016. [see pages 6, 85, 86, 87, 88, 89, 90, 91, 103, 110, and 113]

[Age96]   A.A. Ageev. A triangle-free circle graph with chromatic number 5. *Discrete Math.*, 152(1):295–298, 1996. [see pages 84 and 113]

[ALPS01]   N. Alon, H. Last, R. Pinchasi, and M. Sharir. On the complexity of arrangements of circles in the plane. *Discrete Comput. Geom.*, 26(4):465–492, 2001. [see page 39]

[AP13]   Eyal Ackerman and Rom Pinchasi. On the degenerate crossing number. *Discrete Comput. Geom.*, 49(3):695–702, 2013. [see page 86]

[Ard15]   Federico Ardila. Algebraic and geometric methods in enumerative combinatorics. In Miklós Bóna, editor, *Handbook of Enumerative Combinatorics*, chapter 1.4. CRC Press LLC, Boca Raton, FL, USA, 2015. [see page 109]

[AT07]   Eyal Ackerman and Gábor Tardos. On the maximum number of edges in quasi-planar graphs. *J. Combin. Theory, Ser. A*, 114(3):563–571, 2007. [see pages 55, 58, and 71]

[BB72]   Umberto Bertelè and Francesco Brioschi. *Nonserial dynamic programming*, volume 91 of *Mathematics in Science & Engineering*. Academic Press, New York, 1972. [see page 10]

[BBH⁺17]   Christian Bachmaier, Franz J. Brandenburg, Kathrin Hanauer, Daniel Neuwirth, and Josef Reislhuber. NIC-planar graphs. *Discrete Appl. Math.*, 232:23–40, 2017. [see pages 56 and 67]

[BDE⁺16]   Franz J. Brandenburg, Walter Didimo, William S. Evans, Philipp Kindermann, Giuseppe Liotta, and Fabrizio Montecchiani. Recognizing and drawing IC-planar graphs. *Theoretical Computer Science*, 636:1–16, 2016. [see page 67]

[BDL⁺17]   Michael A. Bekos, Walter Didimo, Giuseppe Liotta, Saeed Mehrabi, and Fabrizio Montecchiani. On RAC drawings of 1-planar graphs. *Theoretical Comput. Sci.*, 689:48–57, 2017. [see page 67]

[BE18]   Michael J. Bannister and David Eppstein. Crossing minimization for 1-page and 2-page drawings of graphs with bounded treewidth. *J. Graph Algorithms Appl.*, 22(4):577–606, 2018. [see pages 72, 88, 92, and 100]

[BEG11]   David A. Brannan, Matthew F. Esplen, and Jeremy J. Gray. *Geometry*. Cambridge Univ. Press, 2nd edition, 2011. [see page 23]

[BGHL18]   Carla Binucci, Emilio Di Giacomo, Md. Iqbal Hossain, and Giuseppe Liotta. 1-page and 2-page drawings with bounded number of crossings per edge. *Eur. J. Comb.*, 68(Supplement C):24–37, 2018. Combinatorial Algorithms, Dedicated to the Memory of Mirka Miller. [see page 71]

[BK79]   Frank Bernhart and Paul C Kainen. The book thickness of a graph. *J. Combin. Theory Ser. B*, 27(3):320–331, 1979. [see page 24]

[BK98]   Heinz Breu and David G. Kirkpatrick. Unit disk graph recognition is NP-hard. *Comput. Geom. Theory Appl.*, 9(1-2):3–24, 1998. [see page 40]

[BKK⁺20]   Michael A. Bekos, Michael Kaufmann, Fabian Klute, Sergey Pupyrev, Chrysanthi Raftopoulou, and Torsten Ueckerdt. Four pages are indeed necessary for planar graphs, 2020. [see pages 13, 83, and 113]

[BKN16]   Jasine Babu, Areej Khoury, and Ilan Newman. Every property of outerplanar graphs is testable. In Klaus Jansen, Claire Mathieu, José D. P. Rolim, and Chris Umans, editors, *APPROX/RANDOM 2016*, volume 60 of *LIPIcs*, pages 21:1–21:19, Dagstuhl, 2016. Schloss Dagstuhl, Leibniz-Zentrum für Informatik. [see page 71]

[Bod96]   Hans L. Bodlaender. A linear-time algorithm for finding tree-decompositions of small treewidth. *SIAM J. Comput.*, 25(6):1305–1317, 1996. [see pages 94 and 99]

[Bor84]   O. V. Borodin. Solution of the Ringel problem on vertex-face coloring of planar graphs and coloring of 1-planar graphs. *Metody Diskret. Analiz.*, 41:12–26, 108, 1984. [see page 71]

[CCJ90]   Brent N. Clark, Charles J. Colbourn, and David S. Johnson. Unit disk graphs. *Discrete Mathematics*, 86(1–3):165–177, 1990. [see page 40]

[CDGK01]   C. C. Cheng, Christian A. Duncan, Michael T. Goodrich, and Stephen G. Kobourov. Drawing planar graphs with circular arcs. *Discrete Comput. Geom.*, 25:405–418, 2001. [see page 2]

[CE12]   Bruno Courcelle and Joost Engelfriet. *Graph Structure and Monadic Second-Order Logic: A Language-Theoretic Approach*. Cambridge Univ. Press, 2012. [see pages 16, 17, 95, 99, and 101]

[CFFP11]   Márcia R. Cerioli, Luérbio Faria, Talita O. Ferreira, and Fábio Protti. A note on maximum independent sets and minimum clique partitions in unit disk graphs and penny graphs: complexity and approximation. *RAIRO Theor. Inf. Appl.*, 45(3):331–346, 2011. [see page 40]

Bibliography

[CFK+15]   Marek Cygan, Fedor V. Fomin, Łukasz Kowalik, Daniel Lokshtanov, Dániel
             Marx, Marcin Pilipczuk, Michał Pilipczuk, and Saket Saurabh. *Parameterized
             Algorithms*, chapter Lower Bounds Based on the Exponential-Time Hypothesis,
             pages 467–521. Springer-Verlag, 2015. [see page 16]

[CFKW19]   Steven Chaplick, Henry Förster, Myroslav Kryven, and Alexander Wolff. On
             arrangements of orthogonal circles. In Daniel Archambault and Csaba D. Tóth,
             editors, *Proc. Graph Drawing and Network Visualization (GD'19)*, volume 11904
             of *Lecture Notes Comput. Sci.*, pages 216–229. Springer, 2019. [see pages 4 and 61]

[CFKW20]   Steven Chaplick, Henry Förster, Myroslav Kryven, and Alexander Wolff. Draw-
             ing graphs with circular arcs and right angle crossings. In Susanne Albers, editor,
             17th Scandinavian Symposium and Workshops on Algorithm Theory (SWAT 2020),
             volume 162 of *LIPIcs*. Schloss Dagstuhl – Leibniz-Zentrum für Informatik, 2020.
             To appear. [see page 4]

[CFL+16]   Steven Chaplick, Krzysztof Fleszar, Fabian Lipp, Alexander Ravsky, Oleg Verbit-
             sky, and Alexander Wolff. Drawing graphs on few lines and few planes. In Yifan
             Hu and Martin Nöllenburg, editors, *Proc. 24th Int. Symp. Graph Drawing &
             Network Vis. (GD'16)*, volume 9801 of *Lecture Notes Comput. Sci.*, pages 166–180.
             Springer-Verlag, 2016. [see pages 3, 13, 21, 22, 23, 24, 25, 26, 27, 32, 33, 36,
             and 37]

[CFL+17]   Steven Chaplick, Krzysztof Fleszar, Fabian Lipp, Alexander Ravsky, Oleg Ver-
             bitsky, and Alexander Wolff. The complexity of drawing graphs on few lines
             and few planes. In Faith Ellen, Antonina Kolokolova, and Jörg-Rüdiger Sack,
             editors, *Proc. Algorithms Data Struct. Symp. (WADS'17)*, volume 10389 of *Lecture
             Notes Comput. Sci.*, pages 265–276. Springer-Verlag, 2017. [see page 22]

[CKL+18]   Steven Chaplick, Myroslav Kryven, Giuseppe Liotta, Andre Löffler, and Alexan-
             der Wolff. Beyond outerplanarity. In Fabrizio Frati and Kwan-Liu Ma, editors,
             GD, volume 10692 of *Lecture Notes Comput. Sci.*, pages 546–559. Springer-Verlag,
             2018. [see page 6]

[CLRS09]   Thomas H. Cormen, Charles E. Leiserson, Ronald L. Rivest, and Clifford Stein.
             Introduction to Algorithms. MIT Press and McGraw-Hill, 3rd edition, 2009.
             [see page 14]

[CLWZ19]   Steven Chaplick, Fabian Lipp, Alexander Wolff, and Johannes Zink. Compact
             drawings of 1-planar graphs with right-angle crossings and few bends. *Comput.
             Geom.*, 84:50–68, 2019. Special issue on EuroCG 2018. [see page 67]

[Cou90]   Bruno Courcelle. The monadic second-order logic of graphs. I. Recognizable
             sets of finite graphs. *Inform. Comput.*, 85(1):12–75, 1990. [see pages 17 and 88]

[CP92]   Vasilis Capoyleas and János Pach. A Turán-type theorem on chords of a convex
             polygon. *J. Combin. Theory Ser. B*, 56(1):9–15, 1992. [see pages 72 and 84]

[CvDK⁺19]   Steven Chaplick, Thomas C. van Dijk, Myroslav Kryven, Ji-Won Park, Alexander Ravsky, and Alexander Wolff. Bundled crossings revisited. In Daniel Archambault and Csaba D. Tóth, editors, *GD*, volume 11904 of *Lecture Notes Comput. Sci.*, pages 63–77. Springer-Verlag, 2019. [see page 7]

[CZQ⁺08]   Weiwei Cui, Hong Zhou, Huamin Qu, Pak Chung Wong, and Xiaoming Li. Geometry-based edge clustering for graph visualization. *IEEE Trans. Vis. Comput. Graph.*, 14(6):1277–1284, 2008. [see page 85]

[de 17]   Éric Colin de Verdière. Computational topology of graphs on surfaces. In Csaba D. Tóth, Joseph O'Rourke, and Jacob E. Goodman, editors, *Handbook of Discrete and Computational Geometry*, chapter 23. CRC Press LLC, Boca Raton, FL, USA, 3rd edition, 2017. [see page 89]

[DEL11]   Walter Didimo, Peter Eades, and Giuseppe Liotta. Drawing graphs with right angle crossings. *Theoret. Comput. Sci.*, 412(39):5156–5166, 2011. [see pages 4, 55, and 66]

[DESW07]   Vida Dujmović, David Eppstein, Matthew Suderman, and David Wood. Drawings of planar graphs with few slopes and segments. *Comput. Geom. Theory Appl.*, 38:194–212, 2007. [see pages 13, 21, 27, 35, and 36]

[DETT99]   Giuseppe Di Battista, Peter Eades, Roberto Tamassia, and Ioannis G. Tollis. *Graph Drawing: Algorithms for the Visualization of Graphs*. Prentice Hall, Upper Saddle River, NJ, 1999. [see pages 11, 15, 33, 40, 50, and 52]

[DEW17]   Vida Dujmović, David Eppstein, and David R. Wood. Structure of graphs with locally restricted crossings. *SIAM J. Discrete Math.*, 31(2):805–824, 2017. [see page 71]

[DGMW10]   Vida Dujmović, Joachim Gudmundsson, Pat Morin, and Thomas Wolle. Notes on large angle crossing graphs. In A. Potanin and A. Viglas, editors, *Proc. Comput. Australasian Theory Symp. (CATS'10)*, volume 109 of *CRPIT*, pages 19–24. Australian Computer Society, 2010. [see pages 55 and 58]

[DKM02]   Andreas W. M. Dress, Jack H. Koolen, and Vincent Moulton. On line arrangements in the hyperbolic plane. *Eur. J. Comb.*, 23(5):549–557, 2002. [see page 72]

[DLM18]   Walter Didimo, Giuseppe Liotta, and Fabrizio Montecchiani. A survey on graph drawing beyond planarity. *ACM Computing Surveys*, 52, 04 2018. [see page 2]

[DMNW11]   Stephane Durocher, Debajyoti Mondal, Rahnuma Islam Nishat, and Sue Whitesides. A note on minimum-segment drawings of planar graphs. *J. Graph Algorithms Appl.*, 17:301–328, 2011. [see pages 13 and 21]

[DN14]   Zdeněk Dvořák and Sergey Norin. Treewidth of graphs with balanced separations. ArXiv, 2014. [see pages 11 and 75]

[Dör65]   H. Dörie. *100 Great Problems of Elementary Mathematics: Their History and Solutions*. Dover, New York, 1965. [see page 41]

Bibliography

[DP11]   Adrian Dumitrescu and János Pach. Minimum clique partition in unit disk graphs. *Graphs & Combin.*, 27(3):399–411, 2011. [see page 40]

[Dun11]   Christian A. Duncan. On graph thickness, geometric thickness, and separator theorems. *Comput. Geom. Theory Appl.*, 44(2):95–99, 2011. [see page 24]

[Ede06]   Herbert Edelsbrunner. Lecture notes for Computational Topology (CPS296.1). https://www.cs.duke.edu/courses/fall06/cps296.1/Lectures/sec-III-3.pdf, 2006. [see page 23]

[EL13]   Peter Eades and Giuseppe Liotta. Right angle crossing graphs and 1-planarity. *Discrete Appl. Math.*, 161(7):961–969, 2013. [see page 67]

[Epp18a]   David Eppstein. Circles crossing at equal angles. https://11011110.github.io/blog/2018/12/22/circles-crossing-equal.html, 2018. Accessed: 2019-06-11. [see page 39]

[Epp18b]   David Eppstein. Triangle-free penny graphs: Degeneracy, choosability, and edge count. In Fabrizio Frati and Kwan-Liu Ma, editors, *Proc. Graph Drawing & Network Vis. (GD'17)*, volume 10692 of *Lecture Notes Comput. Sci.* Springer-Verlag, 2018. [see page 40]

[ES18]   Niklas Eén and Niklas Sörensson. The Minisat Page. http://minisat.se/, 2018. Accessed: 2020-06-26. [see page 82]

[EW96]   Peter Eades and Sue Whitesides. The logic engine and the realization problem for nearest neighbor graphs. *Theoretical Computer Science*, 169(1):23–37, 1996. [see page 40]

[Fel97]   Stefan Felsner. On the number of arrangements of pseudolines. *Discrete Comput. Geom.*, 18:257–267, 1997. [see pages 99 and 103]

[Fel04]   Stefan Felsner. *Geometric Graphs and Arrangements: Some Chapters from Combinatorial Geometry*. Vieweg Verlag, 2004. [see page 39]

[FG06]   Jörg Flum and Martin Grohe. *Parametrized Complexity Theory*. Springer-Verlag, 2006. [see page 11]

[FG18]   S. Felsner and J.E. Goodman. Pseudoline arrangements. In J.E. Goodman, J. O'Rourke, and Cs. D. Tóth, editors, *Handbook of Discrete and Computational Geometry*, chapter 5. CRC Press LLC, Boca Raton, FL, USA, 3rd edition, 2018. [see page 105]

[FHSV16]   Martin Fink, John Hershberger, Subhash Suri, and Kevin Verbeek. Bundled crossings in embedded graphs. In Evangelos Kranakis, Gonzalo Navarro, and Edgar Chávez, editors, *LATIN*, volume 9644 of *Lecture Notes Comput. Sci.*, pages 454–468. Springer-Verlag, 2016. [see pages 6, 85, 87, 89, 110, and 113]

[FMR79]   I. S. Filotti, Gary L. Miller, and John Reif. On Determining the Genus of a Graph in $O(v^{O(g)})$ Steps(Preliminary Report). In *Proceedings of the Eleventh Annual ACM Symposium on Theory of Computing*, STOC '79, page 27–37, New York, NY, USA, 1979. Association for Computing Machinery. [see page 14]

[Fox11]   Jacob Fox. Constructing dense graphs with sublinear hadwiger number. *ArXiv*, abs/1108.4953, 2011. [see page 11]

[FP84]   Z. Füredi and I. Palásti. Arrangements of lines with a large number of triangles. *Proc. Amer. Math. Soc.*, 92(4):561–566, 1984. [see page 39]

[FPS13]   Jacob Fox, János Pach, and Andrew Suk. The number of edges in k-quasi-planar graphs. *SIAM J. Discrete Math.*, 27(1):550–561, 2013. [see page 72]

[FPW15]   Martin Fink, Sergey Pupyrev, and Alexander Wolff. Ordering metro lines by block crossings. *J. Graph Algorithms Appl.*, 19(1):111–153, 2015. [see page 85]

[FS18]   Stefan Felsner and Manfred Scheucher. Arrangements of pseudocircles: Triangles and drawings. In Fabrizio Frati and Kwan-Liu Ma, editors, *Proc. Graph Drawing & Network Vis. (GD'17)*, volume 10692 of *Lecture Notes Comput. Sci.*, pages 127–139. Springer, 2018. [see page 39]

[GHNS11]   Emden R. Gansner, Yifan Hu, Stephen North, and Carlos Scheidegger. Multilevel agglomerative edge bundling for visualizing large graphs. In Giuseppe Di Battista, Jean-Daniel Fekete, and Huamin Qu, editors, *PACIFICVIS*, pages 187–194. IEEE, 2011. [see page 85]

[GHS02]   András Gyárfás, Alice Hubenko, and József Solymosi. Large cliques in C_4-free graphs. *Combinatorica*, 22(2):269–274, 2002. [see page 48]

[Gib85]   Alan Gibbons. *Algorithmic Graph Theory*. Camb. Univ. Press, 1985. [see page 9]

[GJ79]   M. Garey and D. Johnson. *Computers and Intractability: A Guide to the Theory of NP-Completeness*. Series of Books in the Mathematical Sciences. W. H. Freeman, first edition edition, 1979. [see pages 14 and 15]

[GJ83]   Michael R. Garey and David Johnson. Crossing number is NP-complete. *SIAM J. Algebr. Discrete Meth.*, 4:312–316, 1983. [see page 5]

[GJGP80]   M. R. Garey, D. S. Johnson, L. Miller Gary, and C. H. Papadimitriou. The complexity of coloring circular arcs and chords. *SIAM J. Alg. Disc. Meth.*, 1(2):216–227, 1980. [see page 84]

[GKT14]   Jesse Geneson, Tanya Khovanova, and Jonathan Tidor. Convex geometric $(k + 2)$-quasiplanar representations of semi-bar k-visibility graphs. *Discrete Math.*, 331:83–88, 2014. [see page 72]

[Gru72]   Branko Gruenbaum. *Arrangements and Spreads*, volume 10 of *CBMS Regional Conf. Ser. Math.* AMS, Providence, RI, U.S.A., 1972. [see pages 37 and 39]

[Har70]   Frank Harary. Covering and packing in graphs I. *Ann. N.Y. Acad. Sci.*, 175:198–205, 1970. [see page 32]

[Har15]   Frank Harary. *A Seminar on Graph Theory*. Dover Publications, New York, 2015. [see page 24]

[HEF⁺14]   C. Hurter, O. Ersoy, S. I. Fabrikant, T. R. Klein, and A. C. Telea. Bundled visualization of dynamicgraph and trail data. *IEEE Trans. Vis. Comput. Graphics*, 20(8):1141–1157, 2014. [see page 85]

[HEH14]   Weidong Huang, Peter Eades, and Seok-Hee Hong. Larger crossing angles make graphs easier to read. *J. Vis. Lang. Comput.*, 25(4):452–465, 2014. [see page 55]

[HEK⁺15]   Seok-Hee Hong, Peter Eades, Naoki Katoh, Giuseppe Liotta, Pascal Schweitzer, and Yusuke Suzuki. A linear-time algorithm for testing outer-1-planarity. *Algorithmica*, 72(4):1033–1054, 2015. [see page 71]

[HET12]   C. Hurter, O. Ersoy, and A. Telea. Graph bundling by kernel density estimation. *Comput. Graph. Forum*, 31:865–874, 2012. [see page 85]

[HHE08]   Weidong Huang, Seok-Hee Hong, and Peter Eades. Effects of crossing angles. In *Proc. IEEE VGTC Pacific Visualization (PacificVis'08)*, pages 41–46, 2008. [see page 55]

[HK01]   Petr Hliněný and Jan Kratochvíl. Representing graphs by disks and balls (a survey of recognition complexity results). *Discrete Math.*, 229(1–3):101–124, 2001. [see page 40]

[HKMS18]   Gregor Hültenschmidt, Philipp Kindermann, Wouter Meulemans, and André Schulz. Drawing planar graphs with few geometric primitives. *J. Graph Algorithms Appl.*, 22(2):357–387, 2018. [see page 21]

[Hli06]   Petr Hliněný. Crossing number is hard for cubic graphs. *Journal of Combinatorial Theory, Series B*, 96(4):455 – 471, 2006. [see page 5]

[HN16]   Seok-Hee Hong and Hiroshi Nagamochi. Testing full outer-2-planarity in linear time. In Ernst W. Mayr, editor, *WG 2016*, volume 9224 of *Lecture Notes Comput. Sci.*, pages 406–421. Springer-Verlag, 2016. [see pages 71, 72, 73, and 80]

[Hol06]   Danny Holten. Hierarchical edge bundles: Visualization of adjacency relations in hierarchical data. *IEEE Trans. Vis. Comput. Graphics*, 12(5):741–748, 2006. [see pages 14 and 85]

[Hua07]   Weidong Huang. Using eye tracking to investigate graph layout effects. In Seok-Hee Hong and Kwan-Liu Ma, editors, *Proc. Asia-Pacific Symp. Visual. (APVIS'07)*, pages 97–100. IEEE, 2007. [see page 55]

[HvKKR14]   Michael Hoffmann, Marc van Kreveld, Vincent Kusters, and Günter Rote. Quality ratios of measures for graph drawing styles. In *Proc. 26th Canadian Conf. Comput. Geom. (CCCG'14)*, pages 33–39, 2014. [see page 21]

[HW17]   Daniel J. Harvey and David R. Wood. Parameters Tied to Treewidth. *Journal of Graph Theory*, 84(4):364–385, 2017. [see page 11]

[IMS17]   Alexander Igamberdiev, Wouter Meulemans, and André Schulz. Drawing planar cubic 3-connected graphs with few segments: Algorithms & experiments. *J. Graph Algorithms Appl.*, 21(4):561–588, 2017. [see page 21]

[IP01]   Russell Impagliazzo and Ramamohan Paturi. On the complexity of k-SAT. *J. Comput. Syst. Sci.*, 62(2):367–375, 2001. [see page 16]

[KK97]   Alexandr Kostochka and Jan Kratochvíl. Covering and coloring polygon-circle graphs. *Discrete Math.*, 163(1):299–305, 1997. [see page 84]

[KLM17]   Stephen G. Kobourov, Giuseppe Liotta, and Fabrizio Montecchiani. An annotated bibliography on 1-planarity. *Computer Science Review*, 25:49–67, 2017. ArXiv: http://arxiv.org/abs/1703.02261. [see page 71]

[KM93]   Alexandr V. Kostochka and Leonid S. Melnikov. On a lower bound for the isoperimetric number of cubic graphs. In *Proc. 3rd Int. Petrozavodsk Conf. Probabilistic Methods in Discrete Mathematics*, pages 251–265. Moskva: TVP; Utrecht: VSP, 1993. [see page 32]

[KM12]   Ross J. Kang and Tobias Müller. Sphere and dot product representations of graphs. *Discrete Comput. Geom.*, 47(3):548–568, 2012. [see page 53]

[KM14]   Ross J. Kang and Tobias Müller. Arrangements of pseudocircles and circles. *Discrete Comput. Geom.*, 51(4):896–925, 2014. [see page 39]

[KMR08]   Ken-ichi Kawarabayashi, Bojan Mohar, and Bruce A. Reed. A simpler linear time algorithm for embedding graphs into an arbitrary surface and the genus of graphs of bounded tree-width. In *FOCS*, pages 771–780. IEEE, 2008. [see pages 16, 89, and 92]

[KMS18]   Philipp Kindermann, Wouter Meulemans, and André Schulz. Experimental analysis of the accessibility of drawings with few segments. *J. Graph Algorithms Appl.*, 22(3):501–518, 2018. [see pages 12 and 21]

[KR07]   Ken-ichi Kawarabayashi and Bruce Reed. Computing crossing number in linear time. In *STOC*, pages 382–390. ACM, 2007. [see page 110]

[KRW19]   Myroslav Kryven, Alexander Ravsky, and Alexander Wolff. Drawing Graphs on Few Circles and Few Spheres. *J. Graph Algorithms Appl.*, 23(2):371–391, 2019. [see page 3]

Bibliography

[KW01]   Michael Kaufmann and Dorothea Wagner. *Drawing Graphs: Methods and Models*, volume 2025 of *Lecture Notes Comput. Sci.* Springer-Verlag, 2001. [see page 11]

[LH03]   Mark Lombardi and Robert Hobbs, editors. *Mark Lombardi: Global Networks.* Independent Curators, 2003. [see page 2]

[LPVV01]   F. Lazarus, M. Pocchiola, G. Vegter, and A. Verroust. Computing a canonical polygonal schema of an orientable triangulated surface. In *SoCG*, pages 80–89. ACM, 2001. [see page 90]

[LW70]   Don R. Lick and Arthur T. White. k-degenerate graphs. *Canadian J. Math.*, 22:1082–1096, 1970. [see page 10]

[Mit79]   Sandra L. Mitchell. Linear algorithms to recognize outerplanar and maximal outerplanar graphs. *Inform. Process. Lett.*, 9(5):229–232, 1979. [see pages 94 and 97]

[MKNF87] Sumio Masuda, Toshinobu Kashiwabara, Kazuo Nakajima, and Toshio Fujisawa. On the NP-completeness of a computer network layout problem. In *Proc. IEEE Int. Symp. Circuits and Systems*, pages 292–295, 1987. [see page 72]

[MNBR13] Debajyoti Mondal, Rahnuma Islam Nishat, Sudip Biswas, and Md. Saidur Rahman. Minimum-segment convex drawings of 3-connected cubic plane graphs. *J. Comb. Opt.*, 25(3):460–480, 2013. [see page 21]

[Moh99]   Bojan Mohar. A linear time algorithm for embedding graphs in an arbitrary surface. *SIAM J. Discrete Math.*, 12(1):6–26, 1999. [see pages 16 and 89]

[Moh09]   Bojan Mohar. The genus crossing number. *ARS Mathematica Contemporanea*, 2(2):157–162, 2009. [see page 86]

[Mos60]   Moscow Mathematical Olympiad, problem no. 78223. http://www.problems. ru/view_problem_details_new.php?id=78223, 1960. In Russian. [see page 29]

[MP15]   Dániel Marx and Michał Pilipczuk. Optimal parameterized algorithms for planar facility location problems using Voronoi diagrams. In Nikhil Bansal and Irene Finocchi, editors, *ESA*, volume 9294 of *Lecture Notes Comput. Sci.*, pages 865–877. Springer, 2015. [see pages 88 and 102]

[Nak00]   Tomoki Nakamigawa. A generalization of diagonal flips in a convex polygon. *Theor. Comput. Sci.*, 235(2):271–282, 2000. [see page 72]

[Nie06]   Rolf Niedermeier. *Invitation to Fixed-Parameter Algorithms.* Oxford University Press, 2006. [see page 14]

[NR04]   Takao Nishizeki and Md. Saidur Rahman. *Planar Graph Drawing*, volume 12 of *Lecture Notes Comput. Sci.* World Sci. Pub., 2004. [see page 11]

[Ogi69]   C. Stanley Ogilvy. *Excursions in Geometry*. Oxford Univ. Press, New York, 1969.
            [see pages 41, 42, 43, 44, 61, and 62]

[PA95]   János Pach and K. Pankaj Agarwal. *Combinatorial Geometry*. Wiley-Interscience
            Series in Discrete Mathematics and Optimization. John Wiley & Sons, 1995.
            [see page 49]

[PCA02]   Helen Purchase, David Carrington, and Jo-Anne Allder. Empirical evaluation
            of aesthetics-based graph layout. *Empirical Software Engineering*, 7, 09 2002.
            [see page 1]

[PHNK13]   Helen C. Purchase, John Hamer, Martin Nöllenburg, and Stephen G. Kobourov.
            On the usability of Lombardi graph drawings. In Walter Didimo and Maurizio
            Patrignani, editors, *Proc. Graph Drawing (GD'12)*, volume 7704 of *Lecture Notes
            Comput. Sci.*, pages 451–462. Springer, 2013. [see page 2]

[Pin02]   Rom Pinchasi. Gallai–Sylvester theorem for pairwise intersecting unit circles.
            Discrete Comput. Geom., 28(4):607–624, 2002. [see page 39]

[PNBH16]   Sergey Pupyrev, Lev Nachmanson, Sergey Bereg, and Alexander E. Holroyd.
            Edge routing with ordered bundles. *Comput. Geom. Theory Appl.*, 52:18–33,
            2016. [see page 85]

[PRT06]   János Pach, Radoš Radoičić, and Géza Tóth. Relaxing planarity for topological
            graphs. In Ervin Győri, Gyula O. H. Katona, László Lovász, and Tamás Fleiner,
            editors, *More Sets, Graphs and Numbers: A Salute to Vera Sós and András Hajnal*,
            pages 285–300. Springer Berlin Heidelberg, 2006. [see page 56]

[PSS96]   J. Pach, F. Shahrokhi, and M. Szegedy. Applications of the crossing number.
            Algorithmica, 16(1):111–117, 1996. [see page 71]

[PT09]   János Pach and Géza Tóth. Degenerate crossing numbers. *Discrete Comput.
            Geom.*, 41(3):376, 2009. [see page 86]

[Pup17]   Sergey Pupyrev. Mixed linear layouts of planar graphs. *CoRR*, abs/1709.00285,
            2017. [see page 83]

[Pur00]   H.C Purchase. Effective information visualisation: a study of graph drawing
            aesthetics and algorithms. *Interacting with Computers*, 13(2):147–162, 2000. [see
            page 1]

[Rin65]   Gerhard Ringel. Ein Sechsfarbenproblem auf der Kugel. *Abhandlungen aus
            dem Mathematischen Seminar der Universität Hamburg*, 29(1):107–117, 1965. [see
            page 71]

[RS84]   Neil Robertson and Paul D Seymour. Graph minors. III. Planar tree-width. *J.
            Combin. Theory Ser. B*, 36(1):49–64, 1984. [see page 10]

[Sch15]   André Schulz. Drawing graphs with few arcs. *J. Graph Algorithms Appl.*,
            19(1):393–412, 2015. [see pages 2, 12, 13, 21, 27, and 28]

Bibliography

[Sch16]   Ursula Scherm. Minimale Überdeckung von Knoten und Kanten in Graphen durch Geraden. Bachelor's Thesis, Institut für Informatik, Universität Würzburg, 2016. [see pages 27 and 28]

[Sch17]   Marcus Schaefer. The graph crossing number and its variants: A survey. *Electr. J. Combin.*, Dynamic Survey DS21, 2017. [see pages 5, 72, and 85]

[SŠ15]   Marcus Schaefer and Daniel Štefankovič. The degenerate crossing number and higher-genus embeddings. In Emilio Di Giacomo and Anna Lubiw, editors, *GD*, volume 9411 of *Lecture Notes Comput. Sci.*, pages 63–74. Springer, 2015. [see page 86]

[Ste26]   Jakob Steiner. Einige Gesetze über die Theilung der Ebene und des Raumes. *Journal für die reine und angewandte Mathematik*, 1:349–364, 1826. [see page 39]

[Tam13]   Roberto Tamassia. *Handbook of Graph Drawing and Visualization*, volume 81 of *Discrete Appl. Math.* Chapman & Hall/CRC, 2013. [see page 11]

[Tho89]   Carsten Thomassen. The graph genus problem is NP-complete. *J. Algorithms*, 10(4):568–576, 1989. [see pages 14, 16, 87, and 89]

[Tru93]   Richard J. Trudeau. *Introduction to Graph Theory*. New York: Dover Pub, 1993. [see page 9]

[Č07]   Jakub Černý. Coloring circle graphs. *Electr. Notes Discrete Math.*, 29:457–461, 2007. EUROCOMB'07. [see page 84]

[vDFF+17]   Thomas C. van Dijk, Martin Fink, Norbert Fischer, Fabian Lipp, Peter Markfelder, Alexander Ravsky, Subhash Suri, and Alexander Wolff. Block crossings in storyline visualizations. *J. Graph Algorithms Appl.*, 21(5):873–913, 2017. [see page 85]

[vDLMW18]   Thomas C. van Dijk, Fabian Lipp, Peter Markfelder, and Alexander Wolff. Computing storylines with few block crossings. In Fabrizio Frati and Kwan-Liu Ma, editors, *GD*, volume 10692 of *Lecture Notes Comput. Sci.*, pages 365–378. Springer-Verlag, 2018. [see page 85]

[Wei19]   Eric W. Weisstein. Gauss–Bonnet formula. http://mathworld.wolfram.com/Gauss-BonnetFormula.html, 2019. Accessed: 2019-07-25. [see page 47]

[Wik16]   Wikipedia. Circle graph — wikipedia, the free encyclopedia, 2016. [Online; accessed 10-June-2017]. [see page 84]

[WPCM02]   Colin Ware, Helen Purchase, Linda Colpoys, and Matthew McGill. Cognitive measurements of graph aesthetics. *Information Visualization*, 1:103–110, 06 2002. [see page 1]

[WT07]   David R. Wood and Jan Arne Telle. Planar decompositions and the crossing number of graphs with an excluded minor. *New York J. Math.*, 13:117–146, 2007. [see pages 71, 75, and 80]

[XRPH12]   Kai Xu, Chris Rooney, Peter Passmore, and Dong-Han Ham. A user study on curved edges in graph visualisation. In Philip Cox, Beryl Plimmer, and Peter Rodgers, editors, *Proc. Theory Appl. Diagrams (DIAGRAMS'10)*, volume 7352 of *Lecture Notes Comput. Sci.*, pages 306–308. Springer, 2012. [see page 2]

[Yan89]   Mihalis Yannakakis. Embedding planar graphs in four pages. *J. Comput. Syst. Sci.*, 38(1):36–67, 1989. [see pages 13, 80, 83, and 113]

[Yan20]   Mihalis Yannakakis. Planar graphs that need four pages. *Journal of Combinatorial Theory, Series B*, 145:241–263, Nov 2020. [see pages 13, 83, and 113]